BackTrack – Testing Wireless Network Security

Secure your wireless networks against attacks, hacks, and intruders with this step-by-step guide

Kevin Cardwell

BIRMINGHAM - MUMBAI

BackTrack – Testing Wireless Network Security

First published: June 2013

Production Reference: 1180613

Published by Packt Publishing Ltd.
Livery Place
35 Livery Street
Birmingham B3 2PB, UK.

ISBN 978-1-78216-406-7

www.packtpub.com

Cover Image by Vivek Sinha (vivek.ratan.sinha@gmail.com)

Credits

Author
Kevin Cardwell

Reviewers
Aaron M. Woody

Abhinav Singh

Arif Jatmoko

Lee Allen

Acquisition Editors
Martin Bell

Erol Staveley

Commissioning Editor
Yogesh Dalvi

Technical Editor
Nitee Shetty

Copy Editors
Brandt D'Mello

Insiya Morbiwala

Alfida Paiva

Laxmi Subramanian

Project Coordinator
Joel Goveya

Proofreader
Clyde Jenkins

Indexer
Tejal R. Soni

Production Coordinator
Nilesh R. Mohite

Cover Work
Nilesh R. Mohite

About the Author

Kevin Cardwell currently works as a freelance consultant and provides consulting services for companies throughout the world. He developed the Strategy and Training Development Plan for the first Government CERT in the country of Oman and also developed the team to man the first Commercial Security Operations Center there. He has worked extensively with banks and financial institutions throughout the Middle East, Africa, Europe, and the UK. He currently provides consultancy services to commercial companies, governments, major banks, and financial institutions across the globe.

About the Reviewers

Aaron M. Woody is a security consultant with over 15 years of experience in Information Technology, with a focus on security. He is a speaker and an active instructor, teaching hacking, forensics, and information security. In addition to this, he has been a technical reviewer on several titles published by *Packt Publishing*. Aaron maintains two blogs: www.n00bpentesting.com and www.datacentricsec.com. Aaron can also be followed on Twitter at @shai_saint.

Aaron is the author of *Enterprise Security: A Data-Centric Approach to Securing the Enterprise, Packt Publishing*.

Abhinav Singh is a young information security specialist from India. He has a keen interest in the field of hacking and network security and has adopted this field as full-time employment. He is the author of *Metasploit Penetration Testing Cookbook, Packt Publishing*, a book dealing with pen-testing using the most widely -used framework. Abhinav's work has been quoted in several portals and technology magazines. He is also an active contributor to the SecurityXploded community. He can be reached by mail at abhinavbom@gmail.com or on Twitter at @abhinavbom.

I would like to thank my grandparents for their blessings, my parents for their support, and my sister for being my perfect doctor.

Arif Jatmoko (MKom, CISSP, CISA, CCSP, CEH) is an IT Security Auditor at Bank Mandiri, Indonesia, and a private pentester for a few government projects. Prior to joining the bank, Arif had spent over 15 years working as a computer security specialist, computer forensicist, and malware analyst. From the early stages of his career, he has been working with top Fortune 500 companies as an IT security officer and has run several pentest projects for government and military institutions.

Now, he is working on a research about protocol reverse- engineering related to application systems within financial transactions such as banking.

Lee Allen is currently the Vulnerability Management Program lead for one of the Fortune 500 countries.

Lee is also the owner of `miDgames.com`, which is dedicated to bridging the gap between learning and fun by providing 3D video games that teach and reinforce complex subjects such as Linux command-line and penetration-testing skills.

Lee Allen is the author of *Advanced Penetration Testing for Highly-Secured Environments: The Ultimate Security Guide*, Packt Publishing.

www.PacktPub.com

Support files, eBooks, discount offers and more

You might want to visit www.PacktPub.com for support files and downloads related to your book.

Did you know that Packt offers eBook versions of every book published, with PDF and ePub files available? You can upgrade to the eBook version at www.PacktPub.com and as a print book customer, you are entitled to a discount on the eBook copy. Get in touch with us at service@packtpub.com for more details.

At www.PacktPub.com, you can also read a collection of free technical articles, sign up for a range of free newsletters and receive exclusive discounts and offers on Packt books and eBooks.

http://PacktLib.PacktPub.com

Do you need instant solutions to your IT questions? PacktLib is Packt's online digital book library. Here, you can access, read and search across Packt's entire library of books.

Why Subscribe?

- Fully searchable across every book published by Packt
- Copy and paste, print and bookmark content
- On demand and accessible via web browser

Free Access for Packt account holders

If you have an account with Packt at www.PacktPub.com, you can use this to access PacktLib today and view nine entirely free books. Simply use your login credentials for immediate access.

This book is dedicated to Loredana for all of her support and understanding during the many nights of research and writing. Thank you.

Table of Contents

Preface

This book is for the reader who wants to understand more about their wireless network, and how to use a software distribution such as BackTrack to be able to survey their wireless environment and select a robust and secure configuration.

What this book covers

Chapter 1, Installing and Configuring BackTrack, shows the reader how to install, configure, and customize BackTrack. At the end of this chapter, the reader will have a working and customized BackTrack application.

Chapter 2, Working with the Wireless Card, shows the reader how to work with the configuration and deal with the sometimes challenging task of getting their wireless card to work within BackTrack. At the end of this chapter, you will have a wireless card that works with the tools within BackTrack.

Chapter 3, Surveying Your Wireless Zone, covers how to use the tools within BackTrack and examines the wireless environment around you. You will learn how to identify wireless networks and determine the characteristics of these networks. At the end of this chapter, you will have a fundamental understanding of the components that are visible when surveying your zone.

Chapter 4, Breaching Wireless Security, introduces the reader to the way in which hackers typically break into networks. Within this chapter, you will get to practice some of the more common types of attacks. At the end of this chapter, you will have seen the technique used to crack WEP and WPA.

Chapter 5, *Securing Your Wireless Network*, shows you how to apply all of the knowledge gained from the previous chapters, and also gives you the opportunity to examine and evaluate the security settings possible for your wireless network. At the end of this chapter, the reader will be able to make the best decisions when it comes to securing their home wireless networks.

Appendix, *Wireless Tools*, lists a number of tools, with a brief explanation of each tool and links to other resources with respect to the tool.

What you need for this book

A computer with a minimum 2 GB of RAM (4 GB is recommended) is needed. You will also need virtualization software products. The book uses the VMware Workstation; but if you are familiar with others, you can use them. You will require the BackTrack distribution. The steps for its installation and configuration are included within the book.

Who this book is for

This book is for anyone who wants to know more about wireless networks and/ or how to secure their wireless networks. The book has been written for readers at a beginner's level, but they should be familiar with networks. For those who have more experience with the software, this book can serve as a refresher and validation of your skill sets.

Conventions

In this book, you will find a number of styles of text that distinguish between different kinds of information. Here are some examples of these styles, and an explanation of their meaning.

Code words in text, database table names, folder names, filenames, file extensions, pathnames, dummy URLs, user input, and Twitter handles are shown as follows: "Once the machine has booted, you will need to log in with `root` and a password of `toor` (root in reverse)."

Any command-line input or output is written as follows:

```
root@bt:~# /etc/init.d/networking start
```

New terms and important words are shown in bold. Words that you see on the screen, in menus or dialog boxes for example, appear in the text like this: "On the **Downloads** page, you will see a drop-down window that you will use to select the version of the distribution that you want to download."

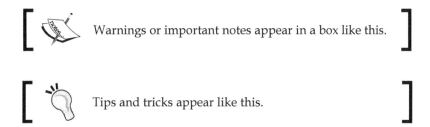

> Warnings or important notes appear in a box like this.

> Tips and tricks appear like this.

Reader feedback

Feedback from our readers is always welcome. Let us know what you think about this book—what you liked or may have disliked. Reader feedback is important for us to develop titles that you really get the most out of.

To send us general feedback, simply send an e-mail to feedback@packtpub.com, and mention the book title via the subject of your message.

If there is a topic that you have expertise in and you are interested in either writing or contributing to a book, see our author guide on www.packtpub.com/authors.

Customer support

Now that you are the proud owner of a Packt book, we have a number of things to help you to get the most from your purchase.

Downloading the example code

You can download the example code files for all Packt books you have purchased from your account at http://www.packtpub.com. If you purchased this book elsewhere, you can visit http://www.packtpub.com/support and register to have the files e-mailed directly to you.

Errata

Although we have taken every care to ensure the accuracy of our content, mistakes do happen. If you find a mistake in one of our books—maybe a mistake in the text or the code—we would be grateful if you would report this to us. By doing so, you can save other readers from frustration and help us improve subsequent versions of this book. If you find any errata, please report them by visiting http://www.packtpub.com/submit-errata, selecting your book, clicking on the **errata submission form** link, and entering the details of your errata. Once your errata are verified, your submission will be accepted and the errata will be uploaded on our website, or added to any list of existing errata, under the Errata section of that title. Any existing errata can be viewed by selecting your title from http://www.packtpub.com/support.

Piracy

Piracy of copyright material on the Internet is an ongoing problem across all media. At Packt, we take the protection of our copyright and licenses very seriously. If you come across any illegal copies of our works, in any form, on the Internet, please provide us with the location address or website name immediately so that we can pursue a remedy.

Please contact us at copyright@packtpub.com with a link to the suspected pirated material.

We appreciate your help in protecting our authors, and our ability to bring you valuable content.

Questions

You can contact us at questions@packtpub.com if you are having a problem with any aspect of the book, and we will do our best to address it.

1
Installing and Configuring BackTrack

In this chapter, we are going to look at the following with respect to using BackTrack:

- Downloading and configuring BackTrack
- Installing BackTrack
- Updating BackTrack
- Validating the interfaces
- Customizing Gnome
- Creating a virtual machine

Before we focus on downloading, installing and configuring Backtrack, I will provide a brief introduction to Backtrack. The BackTrack distribution is actually a combination of two different distributions merged together. There are many distributions that are available, but BackTrack distribution has been created specifically for professional security and penetration testing. The BackTrack distribution was the result of a merger between the two distributions Whax and Auditor. For a brief on some of the distributions, including Auditor, you can find a presentation that I gave at Black Hat in 2005 via this link:

```
http://www.blackhat.com/presentations/bh-usa-05/bh-us-05-cardwell.pdf
```

The name BackTrack comes from the term *backtracking*, which is the name of a search algorithm.

Downloading and configuring BackTrack

The BackTrack distribution comes in a variety of formats; the format you choose is largely a matter of personal preference and your comfort with creating virtual machines. Each one of the distributions has the same tools; they only differ in the utilities that are available with the desktop once you enter the windowing environment. The available distributions come in either a 32-bit or 64-bit distribution; most users will be fine with 32-bit distributions. The main consideration is the amount of random accessory memory (RAM) you will have available for the installation. If you can allocate more than 4 GB of RAM, you can choose the 64-bit distribution; if not then go with the 32-bit option. BackTrack will work with as little as 1 GB of RAM, but the tools you use within it might need more, so 2 GB or more is recommended.

An important point to note is that you should always verify the image files of anything you download. That is why there is an MD5 hash for each of the downloads. If you do not know how to do this, you can do a search for it on the Internet. There are a number of tools that can be downloaded to assist you with verifying the hash of a file.

The ISO file allows you to take the image and burn it to a DVD, allowing you to boot from the image and run BackTrack from the DVD. Alternatively, you can also mount the image using a virtualization tool and boot it that way.

For our purposes, we will use the virtual machine, as it is easy to use and has been configured with the tools; this allows you to copy and paste within the VM and have a full screen virtual environment. We will also include steps later in the chapter for booting the ISO image in VMware Workstation, as that will result in the identical interface that will be experienced when booting from a DVD image. VMware Workstation was one of the first virtual environment software products and allows us to run multiple computers on one machine.

BackTrack can be downloaded from its official website at `http://www.BackTrack-linux.org/downloads/`. Once you go to the website, you will see that there is a registration request there; this is optional, and downloading the distribution is not required. On the **Downloads** page, you will see a drop-down window that you will use to select the version of the distribution that you want to download. Once you click on the drop-down window, the other windows will be populated as displayed in the next screenshot:

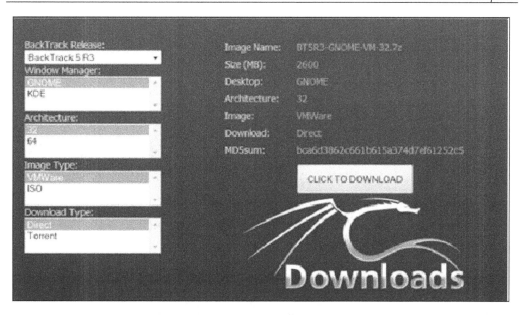

We will be using the BackTrack 5 R3 distribution with the window manager Gnome
(32-bit) and the VMware image installation for the book; download the image, and
then we will continue with the installation.

Installing BackTrack

Once you have downloaded the image, you will need to have certain VMware
software to run the image; you can get the software from www.vmware.com. VMware
Player is smaller with fewer options, although VMware Workstation contains many
features. Both versions have a trial version available, and it is recommended that you
select one and try it out. For this book, we are using VMware Workstation Version 9
as that is the latest version as of this writing. To access the software downloads page,
go to http://www.vmware.com/products/workstation/index.html. VMware
software allows you to perform many functions in a virtual environment, and this
can be an additional security measure for you. If you are using a Mac, you will need
to download VMware Fusion for that; refer to http://www.vmware.com/products/
fusion/overview.html. Also, if you want to use the ISO image to create a DVD and
boot from that, you can do that too. The way in which we start BackTrack is the only
thing that differs from the other distributions; once they start, they all require the
same number of steps to be configured and used.

You may also want to note here that there are more virtualization products than VMware, and if you want to try others, you may; some of the popular, free ones are:

- VirtualBox – `https://www.virtualbox.org/wiki/Downloads`
- Xen – `http://www.xen.org/products/downloads.html`
- Hyper-V – `http://www.microsoft.com/en-us/server-cloud/hyper-v-server/default.aspx`

Hyper-V requires you to have either Windows Server 2008 or Windows Server 2012 installed. There are two reasons why we are using VMware Workstation in this book:

- It has fantastic documentation and support
- The image from BackTrack has the tools installed, and this makes for much better user experience

As always, software is a matter of personal preference and taste, so try different versions and see which one works best for you.

The procedure for starting the tool once you have downloaded the virtual machine is pretty straightforward. Open the image in the virtual machine tool you are using and check the settings. You should see a screen similar to the one shown in the next screenshot. You want to check that there is a network adapter present, and that the RAM is set to a value that you can support on your host machine. As a reminder, 2 GB of RAM is recommended. An example is shown in the next image where the RAM is not sufficient and would need to be changed; you do this by clicking on **Edit virtual machine settings**. Once you have configured this, click on **Power on this virtual machine**. If you are prompted before the VM boots up, leave the default settings, as they are and let the machine boot.

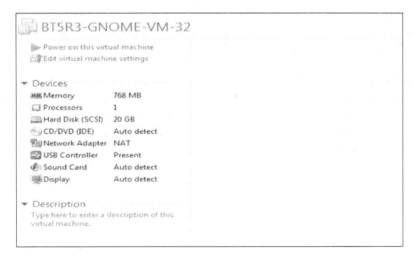

Once the machine has booted, you will need to log in with `root` as the ID and a password of `toor` (root in reverse). This will bring you to the shell interface of BackTrack. We want to start up the windowing environment, but before we do that, we want to check and verify that our network has started. Enter `ifconfig`; you should see a result similar to in the next screenshot:

In some versions of the BackTrack distribution, you might not have an interface named `eth1`. If you only have the `lo` (loopback) interface, you have to start the network; the way to do this is by using `/etc/init.d/networking start`. This will start your network.

```
root@bt:~# /etc/init.d/networking start
```

To avoid having to do this the next time you start BackTrack, enter the `update-rc.d networking defaults` command. This will change the configuration, and each time you start BackTrack, the network will be started for you.

```
root@bt:~# update-rc.d networking defaults
```

There is a possibility that when you restart, you will not get a network address; this is rare with the latest software, but just in case, you might want to enter the `ifconfig` command again.

```
root@bt:~# ifconfig
```

If you are not connected to the network, that is, if you don't have an IP address, you can enter `dhclient`.

```
root@bt:~# dhclient
```

Next, you want to start the XWindows environment; you do this by entering the startx command in the command shell.

```
root@bt:~# startx
```

This will start the XWindows Gnome desktop. Once the window comes up, the first thing you want to do is to open a shell on your desktop. You can accomplish this by going to **Applications | Accessories | Terminal**. The next thing you want to do is to fix your display within the virtual environment inw the VMware Workstation menu at the top of the screen. Navigate to **View | Fit Guest Now** to correct the display; your screen should now be in fullscreen mode, as shown in the following screenshot:

Next, you want to change the password from the default value of toor, which is not a strong password. And to be honest, everyone knows it; so let us change it now. In your command shell terminal window, enter passwd.

```
root@bt:~# passwd
```

This will start the password change process. Enter a password of your choice; it will not echo on the screen as you type it in, and you will be requested to re-enter your password to confirm it. Make sure you do not forget your password, as it can be a challenge to recover it; it is not impossible, but it is a challenge nonetheless and is beyond the scope of this book.

As you can see from the previous screenshot, there is a transparent background in the shell. There is a simple procedure to customize your shell, and we will do that now. I prefer to have a white background with black text, which we will now configure. In the title of the terminal window, navigate to **Edit | Profile Preferences**. Once the window opens, you want to click on **Background**; this is where you will set the transparency of the shell window. My preference is to have no transparency, click on **Transparent Background**, then drag the slider to the right and set it to **Maximum** as shown in the next screenshot:

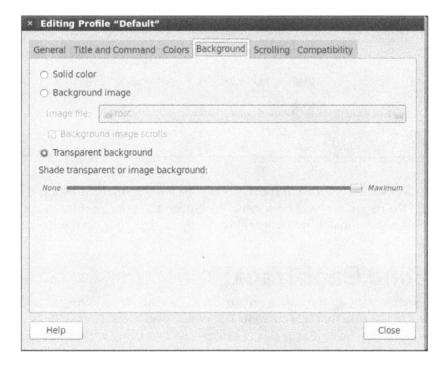

The next thing we will set is the actual colors of the terminal windows. We do this by clicking on **Colors** and then setting the colors for the terminal as identified in the following screenshot:

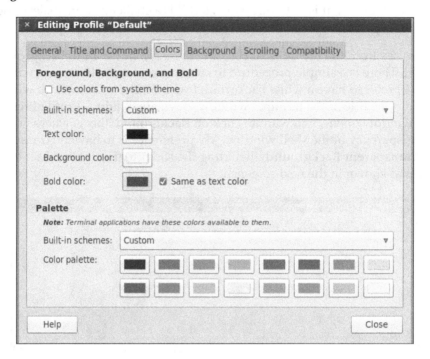

Once you have set the colors and transparency for the terminal, you should have a shell with the colors that you prefer and customized in your profile, so that each time you open a terminal window it will be configured to your preferences.

Updating BackTrack

Now that we have set up the basics on our BackTrack machine, we want to update the tool and make sure that it is current. The developers of BackTrack have instituted an update capability using the `apt-get` utility. The first thing we want to do is to update the package. This requires a working Internet connection on your host; verify this by opening a web browser window and accessing a website. Once you have verified that you are connected to the Internet, go back to the BackTrack machine and enter the `apt-get update` command in a terminal window.

```
root@bt:~# apt-get update
```

Once you have entered the command, you should see a display similar to the one shown in the next screenshot:

```
root@bt:~# apt-get update
Hit http://32.repository.backtrack-linux.org revolution Release.gpg
Hit http://all.repository.backtrack-linux.org revolution Release.gpg
Ign http://32.repository.backtrack-linux.org/ revolution/main Translation-en_US
Ign http://32.repository.backtrack-linux.org/ revolution/microverse Translation-en_US
Ign http://all.repository.backtrack-linux.org/ revolution/main Translation-en_US
Ign http://all.repository.backtrack-linux.org/ revolution/microverse Translation-en_US
Ign http://32.repository.backtrack-linux.org/ revolution/non-free Translation-en_US
Ign http://32.repository.backtrack-linux.org/ revolution/testing Translation-en_US
Hit http://32.repository.backtrack-linux.org revolution Release
Ign http://all.repository.backtrack-linux.org/ revolution/non-free Translation-en_US
Ign http://all.repository.backtrack-linux.org/ revolution/testing Translation-en_US
Hit http://32.repository.backtrack-linux.org revolution/main Packages
Hit http://all.repository.backtrack-linux.org revolution Release
Hit http://32.repository.backtrack-linux.org revolution/microverse Packages
Hit http://all.repository.backtrack-linux.org revolution/main Packages
Hit http://32.repository.backtrack-linux.org revolution/non-free Packages
Hit http://32.repository.backtrack-linux.org revolution/testing Packages
Hit http://all.repository.backtrack-linux.org revolution/microverse Packages
Hit http://all.repository.backtrack-linux.org revolution/non-free Packages
Hit http://all.repository.backtrack-linux.org revolution/testing Packages
Hit http://source.repository.backtrack-linux.org revolution Release.gpg
Ign http://source.repository.backtrack-linux.org/ revolution/main Translation-en_US
Ign http://source.repository.backtrack-linux.org/ revolution/microverse Translation-en_US
Ign http://source.repository.backtrack-linux.org/ revolution/non-free Translation-en_US
Ign http://source.repository.backtrack-linux.org/ revolution/testing Translation-en_US
Hit http://source.repository.backtrack-linux.org revolution Release
```

Once the update has finished, there should be a message saying that the update was successful. Once the update is complete, the next thing to do is to upgrade the distribution itself; we do that also with the apt-get command. Enter apt-get dist-upgrade in the terminal window.

root@bt:~# apt-get dist-upgrade

The output of this command should be similar to the one in the following screenshot:

```
root@bt:~# apt-get dist-upgrade
Reading package lists... Done
Building dependency tree
Reading state information... Done
Calculating upgrade... Done
The following NEW packages will be installed:
  libdevice-serialport-perl
The following packages will be upgraded:
  android-sdk backtrack-utils bbqsql beef bluez-hcidump burpsuite easy-creds
  impacket-examples intercepter-ng jigsaw johnny magictree mantra mercury reaver rfidiot
  se-toolkit skipfish smartphone-pentest-framework subterfuge tcpdump tcpflow thc-ipv6
  uberharvest uniscan volatility watobo websploit wifite wireshark
30 upgraded, 1 newly installed, 0 to remove and 0 not upgraded.
Need to get 219MB of archives.
After this operation, 319kB of additional disk space will be used.
Do you want to continue [Y/n]?
```

The output indicates the packages that are available and will be upgraded. Once you have reviewed them, you are ready to do the upgrade; enter Y to start the upgrade. Once the upgrade has finished, you are done with the main updates and upgrades for BackTrack.

Validating the interfaces

There is one more thing we want to correct when it comes to our network, because there is a problem with the allocation of the interfaces at boot time when you make copies of a Linux virtual machine. If you enter `ifconfig` and your interface does not say `eth0`, you should correct the network.

```
root@bt:~# ifconfig
```

We do this by opening the configuration file and removing all references to the network interfaces that are there. In your shell, enter `gedit`.

```
root@bt:~# gedit
```

This will open the graphical editor that is contained within the Gnome distribution. In the editor, navigate to **File | Open**; when the window opens, navigate to the configuration file that is located at `/etc/udev/rules.d/`. To navigate to the folder, click on the `File System` folder in the window. This will put you in the root of the file system; then, it is only a matter of drilling down until you get into the `rules.d` folder. Select and open the `70-persistent-net.rules` file. Once you are in the file, delete the old interfaces that are there; highlight the line that starts with `# PCI Device` and all the lines after that right up to the end of the file, and then delete them as shown in the next screenshot:

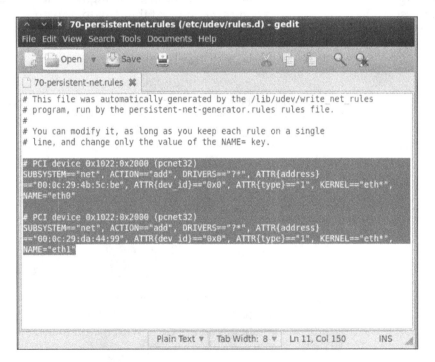

Once you have deleted the interface references, save the file, and quit the program. Click on **Save** and then **Quit**. This will return you to the terminal window and your shell. Now, enter `reboot` and let the machine reboot and regenerate your interfaces.

Once the machine reboots, log in to the machine with the username `root` and whatever password you set earlier (you do remember it, right?). Once you are logged in, start XWindows by entering the `startx` command.

root@bt:~# startx

Once XWindows has started, open the terminal window by using the shortcut on the toolbar or by navigating to **Applications | Accessories | Terminal**. If you have not done so already, correct your display by navigating to **Edit | Fit Guest Now** in VMware Workstation.

We now want to check and verify that our network interfaces are correct; we do this by using our `ifconfig` command. Enter `ifconfig` in the terminal window. As we have removed the extra information that was in the `70-persistent-net.rules` file, we should now see that our interface is `eth0` and not something else; this is shown in the next screenshot. For some reason, when you copy virtual machines, there is a possibility that the programmer did an append to the file and did not overwrite the entries that are in the file; if you make three or four copies of the virtual machine, it will show `eth5` or `eth6`. Again, we will have to correct it by deleting the entries that are in the `70-persistent-net.rules` file. This is the same process for many of the Linux virtual machines when you copy them.

```
root@bt:~# ifconfig
eth0      Link encap:Ethernet  HWaddr 00:0c:29:da:44:99
          inet addr:192.168.177.141  Bcast:192.168.177.255  Mask:255.255.255.0
          inet6 addr: fe80::20c:29ff:feda:4499/64 Scope:Link
          UP BROADCAST RUNNING MULTICAST  MTU:1500  Metric:1
          RX packets:76 errors:0 dropped:0 overruns:0 frame:0
          TX packets:18 errors:0 dropped:0 overruns:0 carrier:0
          collisions:0 txqueuelen:1000
          RX bytes:7600 (7.6 KB)  TX bytes:1918 (1.9 KB)
          Interrupt:19 Base address:0x2000

lo        Link encap:Local Loopback
          inet addr:127.0.0.1  Mask:255.0.0.0
          inet6 addr: ::1/128 Scope:Host
          UP LOOPBACK RUNNING  MTU:16436  Metric:1
          RX packets:40 errors:0 dropped:0 overruns:0 frame:0
          TX packets:40 errors:0 dropped:0 overruns:0 carrier:0
          collisions:0 txqueuelen:0
          RX bytes:8677 (8.6 KB)  TX bytes:8677 (8.6 KB)

root@bt:~#
```

Customizing Gnome

We now have a working, functional, and updated instance BackTrack. Remember to update and upgrade the tool often. We are using the Gnome desktop, we will now discuss some of the methods of customizing the Gnome desktop to suit our tastes. You can skip this section if you are happy with the default desktop or if you have configured Gnome before. For those of you who want to personalize and change the dark look of the Gnome desktop, this section is for you.

There are three main areas we will consider when customizing the Gnome desktop. They are:

- Background
- Theme
- Fonts

To change your background, navigate to **System | Preferences | Appearance | Background**. As you can see, there are not many background types stored by default, so click on **Get more backgrounds online**. You now have many backgrounds to choose from. I like to visit island locations, so I am going to select the **Fakarava Coconut Tree** background. Once you have selected your background, right-click on it and select **Set as Desktop Background**. You will notice that when you open it, the image does not occupy the entire screen; to make the image fullscreen, select the drop-down menu item **stretch**, or select **tile** to have multiple copies of the image displayed on the desktop. The resolution can present challenges, so use **tile** if the resolution does not match the Gnome settings on your machine. After you have finished configuring the settings, click on **Set as Desktop Background**. The next screenshot shows my desktop with the **tile** setting displayed:

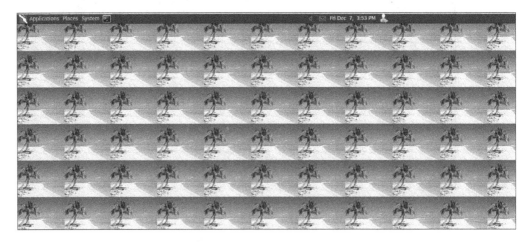

The next thing we want to customize is our theme; we do this the same way we customized the background. Navigate to **System | Preferences | Appearance | Theme**.

This time, we have three themes to choose from instead of the two stock backgrounds; we also have the option to get more themes online, but for our purposes, the **New Wave** option is the one we will use. Click on **New Wave**, and then click on **Close**. If you refer to the next screenshot, we now have less of a "dark" look to our menu items. Again, this is largely a matter of personal preference, and you should experiment with the different options and select the one that works for you.

New wave theme

Now it is time to look at changing the fonts for our desktop. We access the font setting by navigating to **System | Preferences | Appearance | Fonts**. As you can see, there are many fonts from which to choose; again, it is a matter of personal preference and taste. The options are shown in the following screenshot:

Creating a virtual machine

Before we wrap up this chapter, I am sure some of the readers would like to create their own virtual machine or even boot from a DVD. This section has been included for them. I am also one of those who would choose to create their own virtual machine rather than download a prebuilt one. As you will discover, when you take this route, it works well until you attempt to install tools within the virtual machine. As of this writing, there is no easy procedure for this, and it takes quite a lot of time to accomplish it using the current methods that are available.

To get started, the first thing we have to do is to create a virtual machine. For this, we use the virtual machine wizard within VMware Workstation by navigating to **File | New Virtual Machine**. This will start the virtual machine wizard; leave the default setting of **Typical**, and click on **Next**. On the next window, we want to direct the wizard to our ISO image. So, click on the Installer disc image file (the ISO file), and then browse to the location and select the ISO image you downloaded for BackTrack, as shown in the following screenshot:

Once you have selected the ISO image, you will notice that the wizard cannot detect the operating system; we will have to specify it. This is normal behavior and nothing to be alarmed about; after reading the message, click on **Next**.

The operating system selection should be set to Linux by default. You just need to select the type of Linux operating system by clicking on the drop-down arrow and selecting the **Ubuntu** option since this is what the Black Hat distribution is based. Once you click on **Next**, you will have the option to name your virtual machine and the location to store it in. If you would like to change the name, this is the place to do that. Once you have set the name and location you want to use, click on **Next**. Leave the size and other settings at their default values, and click on **Next**. On the next screen, you have a chance to review your hardware settings, and if need be, to customize them. If you are comfortable with the default RAM of 1 GB, click on **Finish**. The following screenshot shows what your hardware should look like:

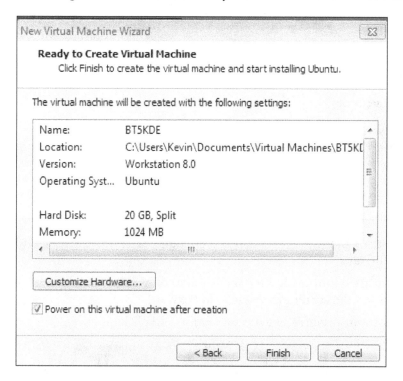

Once you click on **Finish**, the wizard will boot the virtual machine. Since we have configured it to use the ISO image, it will boot from there. At the prompt, click anywhere on the black area and then press *Enter* to boot BackTrack. Once the OS has booted, you will be automatically logged in; for reference, remember that the username is root and the password is toor, and you should change it after installation.

If you want to verify your network settings, enter `ifconfig`; as you will see, the interface is properly set at `eth0` since we booted this from the ISO image and did not copy it. This is reflected in the following screenshot:

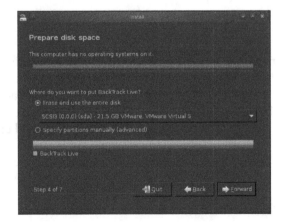

In the command prompt, enter `startx` to launch the desktop and window environment.

```
root@bt:~# startx
```

Once the desktop has started, you will notice that you now have an icon on the desktop for installing the OS to the hard disk. If you are in a virtual environment, that is the recommended next step.

To install BackTrack to the hard disk, there is a series of steps to be followed:

1. Click on the **Install BackTrack** icon, and start the installation process.
2. When a window comes up, select the language for the install, and click on **Forward**.
3. If you are happy with what the installation wizard finds for the time zone and the current time, click on **Forward**.
4. Select your preferred keyboard layout, and click on **Forward**.

At this point, you are at the "prepare the hard disk for installation" stage; make sure you are in a virtual machine and not a bootable DVD, because this will erase the hard drive! If you are not sure, do not proceed until you are. If you want to make sure, refer to the next screenshot and notice that the message **This computer has no operating systems on it.** is identified:

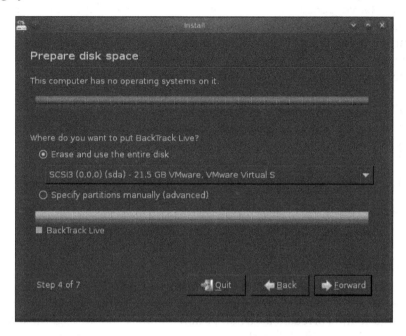

Now that we have covered that and you have verified that you are not going to destroy your system, click on **Forward**.

The installation is now ready to start; this is your last chance to ensure you do not erase your hard drive. If you are comfortable with this, click on **Install**. After some time, the installation will successfully complete (hopefully), and you will see a message like the one in the following screenshot:

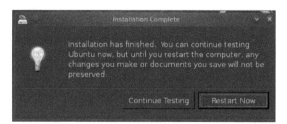

After the installation is complete, you will notice that you cannot adjust the display by selecting the **Fit Guest Now** option because it is grayed out; this is because you do not have the VMware tools installed. You can manually install the tools, but it is not a simple and straightforward procedure. Those who want to try it can access the step-by-step instructions at `http://www.BackTrack-linux.org/wiki/index.php/VMware_Tools`. Again, this process requires updating the kernel and then rebuilding it. The procedure does work, but it is a time-consuming process; only attempt it if you want to go through a challenging process and have a couple of hours to spare. The majority of you will prefer to download the virtual machine that is preconfigured and offered on the BackTrack distribution site.

Downloading the example code

You can download the example code files for all Packt books you have purchased from your account at `http://www.packtpub.com`. If you purchased this book elsewhere, you can visit `http://www.packtpub.com/support` and register to have the files e-mailed directly to you.

Summary

In this chapter we covered a little bit of the history of BackTrack and discussed the different types of installation options. After installing the operating system, we configured it and corrected a virtualization anomaly that caused the interfaces to not be correctly identified. We also covered some standard configuration changes that you can make to your BackTrack desktop to personalize your install. The chapter concluded with how to use the ISO image to create your own virtual machine and customize it and also how to install it to the hard disk.

You should have BackTrack working and configured before we proceed to the next chapter and discuss and select a wireless card for our tools.

2
Working with the Wireless Card

In this chapter, we are going to look at the following topics with respect to using BackTrack:

- Checking card compatibility
- Detecting the wireless card during the boot process
- Detecting the wireless card using `iwconfig`
- Wireless card modes
- Protocol analysis with the wireless card

Now that we have a configured and updated BackTrack, it is time to turn our attention to working with our wireless card. Unfortunately, with the Linux operating system, this is not as straightforward as we would hope. The work that will need to be done depends on a number of factors. Out of all of the factors, the most important one is that Linux goes by the chipset of the card.

Checking card compatibility

There are a number of factors to consider when we work with a wireless card in Linux, and the most important one is to know what the chipset of our wireless card is. This is because we have to know whether we have a driver to support the card, and it takes work to determine what that chipset is. The challenging cards are normally the ones that are built into laptops; in most cases, this card will not work in BackTrack without doing some work and spending some time on it.

The first place to look for information, whether or not your card will work, is on the BackTrack website. There is a listing of supported cards on the BackTrack website at `http://www.BackTrack-linux.org/wiki/index.php/Wireless_Drivers`. The site is shown in the next screenshot:

Getting a wireless card to work can be a challenging process, and it is often time-consuming. However, there are times when you do get lucky and the card is recognized at boot, but more often than not, it takes additional work. When we use virtualization, it is rare for the card to be recognized at boot; this is especially the case with BackTrack. It is not impossible, but we still go through the steps to verify whether there is more work to do. If, however, you are using the boot DVD or another method without the virtual machine, there is a slight chance that your built-in card has been recognized, but again, there is just a chance, so that is why we go through the process.

If you have not already done so, start up your BackTrack machine. Once it starts up, log in with the username `root`, and the password should be whatever you configured it to be. As a reminder, the password is `toor` if you are using a default configuration. After you log in, start the windowing environment by entering `startx`.

```
root@bt:~# startx
```

This should place you in the windowing environment. Your screen should look similar to the one shown in the following screenshot:

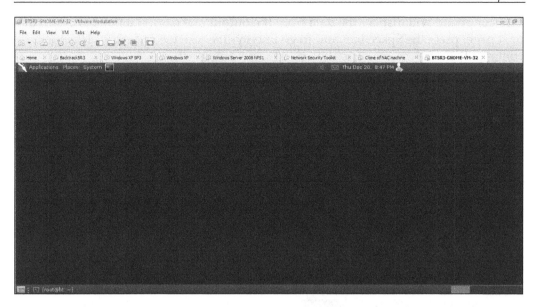

Detecting the wireless card during the boot process

Now that we have the Gnome desktop started, the next thing we will do is open a terminal window. You can do this by clicking on the shortcut icon, or go the long way by navigating to **Application | Accessories | Terminal**. The result is that a terminal window opens. In this terminal window, enter the dmesg command.

```
root@bt:~# dmesg
```

The result of this command will make you scroll your screen a lot. The dmesg command displays the contents of the ring buffer. Do not worry if you do not know what a buffer is, just think of it as a place used to store data. This ring buffer that is displayed contains the messages that the system displays as it reads and encounters the hardware and devices of your machine. To scroll up and view this would be a bit of a challenge, but the one thing to remember about Linux is that they have devised ways for us to do this much more efficiently. The first method we will look at is the more option; enter the following in the terminal window:

```
root@bt:~# dmesg | more
```

This command will display the data from the output one page at a time. To go to the next page, press the Space bar, and to go line-by-line, use the *Enter* key. The more command, used with dmesg, is shown in the next screenshot:

```
root@bt:~# dmesg | more
[    0.000000] Initializing cgroup subsys cpuset
[    0.000000] Initializing cgroup subsys cpu
[    0.000000] Linux version 3.2.6 (root@bt) (gcc version 4.4.3 (Ubuntu 4.4.3-4u
buntu5) ) #1 SMP Fri Feb 17 10:40:05 EST 2012
[    0.000000] KERNEL supported cpus:
[    0.000000]   Intel GenuineIntel
[    0.000000]   AMD AuthenticAMD
[    0.000000]   NSC Geode by NSC
[    0.000000]   Cyrix CyrixInstead
[    0.000000]   Centaur CentaurHauls
[    0.000000]   Transmeta GenuineTMx86
[    0.000000]   Transmeta TransmetaCPU
[    0.000000]   UMC UMC UMC UMC
[    0.000000] Disabled fast string operations
[    0.000000] BIOS-provided physical RAM map:
[    0.000000]  BIOS-e820: 0000000000000000 - 000000000009f400 (usable)
[    0.000000]  BIOS-e820: 000000000009f400 - 00000000000a0000 (reserved)
[    0.000000]  BIOS-e820: 00000000000ca000 - 00000000000cc000 (reserved)
[    0.000000]  BIOS-e820: 00000000000dc000 - 0000000000100000 (reserved)
[    0.000000]  BIOS-e820: 0000000000100000 - 000000002fee0000 (usable)
[    0.000000]  BIOS-e820: 000000002fee0000 - 000000002feff000 (ACPI data)
[    0.000000]  BIOS-e820: 000000002feff000 - 000000002ff00000 (ACPI NVS)
[    0.000000]  BIOS-e820: 000000002ff00000 - 0000000030000000 (usable)
```

Again, this is probably not the best option, so we will look at option two. Option number two is to use the less command by entering the following into the terminal window:

```
root@bt:~# dmesg | less
```

The saying that we use in Linux, "less is more", is because unlike the more command in which the text scrolls on as you go through it, the less command allows you to use the *Page Up*, *Page Down*, and arrow keys without having the page scroll away from you! This is why less really is more.

We still have the challenge of looking through all of this ring buffer stuff to find our wireless card, even if it is there. Fortunately, we have another option that is even better than the other two! We will explore this now. Rather than looking through the output line-by-line, we can use a powerful utility named grep. The grep tool is an essential one within the Linux toolset that is used extensively when searching for a string.

The `grep` tool will search a file for the specified string, and then display the corresponding line(s) containing that string. With respect to `dmesg` and the message for which we are looking, we search for the following two components:

- wireless
- wlan0

The `grep` search queries for these two strings, as shown in the next screenshot:

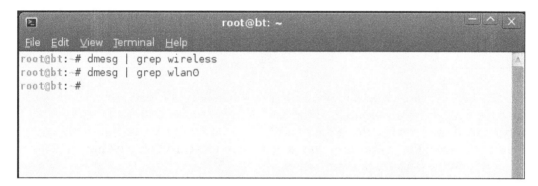

If you have not already done so, enter these two commands, and see if your output matches what is displayed in the previous screenshot. Unfortunately, in most cases, this is what you will see; this means that during the boot process, the wireless device was not detected. There are a number of things we can do to see if we can get the card recognized after the boot, and we will do that in a moment.

Detecting the wireless card using iwconfig

The next thing we want to do is to look at the `iwconfig` tool. In your terminal window, enter `iwconfig`:

```
root@bt:~# iwconfig
```

A sample output of this command is shown in the following screenshot; it shows that there are no wireless cards currently recognized on this machine.

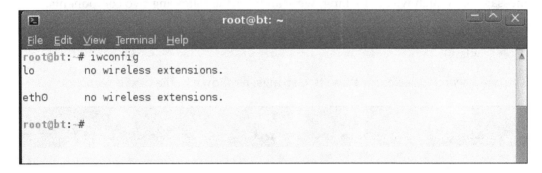

At this time, you are probably wondering what we have to do to get the card recognized? Unfortunately, if the card is not one among those that are recommended, it can become a daunting task. So, we will try a few more things before we reveal the guaranteed way to get a wireless card that we know will work within the BackTrack distribution.

When you are in a virtual environment, you can see whether the device is listed or just not connected. We do this by clicking on **Removable Devices** under the **VM** menu. This will bring up a list of removable devices found on the machine, which is displayed in the following screenshot:

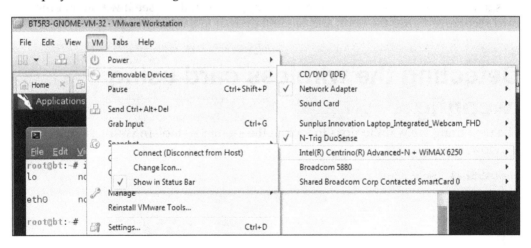

The process now is to connect each device by selecting the available wireless card in the host OS, and then use the `iwconfig` command again, and see if you get lucky; if you do, then that's it! It is more than likely that you will not get lucky with this method. So, in reality, the best solution is to purchase a wireless card that will work with BackTrack, and more importantly, support the features that you will need with the tools. There are two that I highly recommend, both of which are inexpensive. They are the following:

- **Wi-Fire** (www.hfield.com)
- **Alpha AWUS036H** (http://www.alphawireless.com/)

Both of these cards are USB, which means they should work as soon as you put them in; also, as already mentioned, they are inexpensive. The Alpha card will give you 1000 mW of power, which is the highest transmission on signal power that is allowed by the Federal Communications Commission (FCC). The FCC regulates interstate and international communications by radio, television, wire, satellite, and cable in the U.S. and U.S. territories. Other countries may have different regulations.

Configuring the wireless card

I am sure there are some of you who still want to try and get your card working, so we have a few more steps you can try, and we will cover them now. Having said that, it is highly recommended that you use one of the two cards mentioned earlier.

The next thing to try is the `airmon-ng` command. There is very little chance that your card will be here when it is nowhere else, but we can never say never, because this is electronics! In your terminal window, enter `airmon-ng.`, using the following command:

```
root@bt:~# airmon-ng
```

The output of this command is shown in the next screenshot:

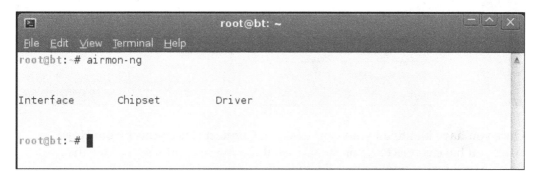

Once again we have failed! The card is not recognized, so now it is a matter of searching the Internet and trying to find a fix to get your card working. One alternative is to use any external card you may have. If you have an external card, plug it in and go through the same steps as before. There are two places to go to try and read more about getting your card operational. They are http://www.BackTrack-linux.org/forums/showthread.php?t=34469 and http://www.youtube.com/watch?v=MxTErUIBlyo. Once again, it is highly recommended that you purchase one of the USB wireless cards mentioned previously. It will save you a lot of time and stress, because even a direct PCI card may not be detected even if the card is supported. A simple search on the Internet can also provide more information when you are having difficulties.

If you have a recognized card or a USB external card, there are several steps to follow to identify the card. For the purposes of this book, the card that is being used is the Wi-Fire card from hField Technologies. The first thing that you need to do is insert the card into the USB port of the computer. Depending on the setup, you may or may not connect to it on the virtual machine. To get the card recognized in BackTrack, we have to connect to it using the VM configuration. We do this by clicking on **Removable Devices** under the **VM** menu and looking for the card listed there. The example in the next screenshot shows that the Wi-Fire card is showing up as **Z-Com USB Device**. If you are using another card, it is a matter of trial and error to determine what it is called, unless you get lucky and the card has a recognizable name (which is rare).

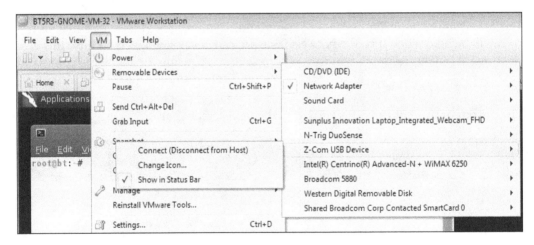

Once you have identified your card, click on **Connect (Disconnect from Host)**. After you have connected, you should see if the dmesg command can see the card. In your terminal window, enter the following command:

```
root@bt:~# dmesg | grep wlan0
```

Hopefully, you will see something similar to what is shown in the next screenshot:

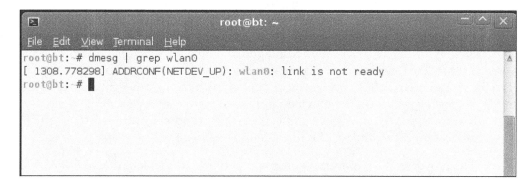

If you do not see any indication of **wlan0**, you can reboot the system and see whether it recognizes the card after the reboot. As a reminder, to reboot the system, enter reboot in a terminal window:

root@bt:~# reboot

Now that you have the **wlan0** device listed in the output of dmesg, you need to start up the wireless card; before you do that, enter the following command into a terminal window:

root@bt:~# iwconfig

You should now see that there is an adaptor that has wireless extensions. The next screenshot shows that the wlan0 interface has wireless extensions:

```
root@bt:~# dmesg | grep wlan0
[ 1308.778298] ADDRCONF(NETDEV_UP): wlan0: link is not ready
root@bt:~# iwconfig
lo        no wireless extensions.

wlan0     IEEE 802.11bg  ESSID:off/any
          Mode:Managed  Access Point: Not-Associated   Tx-Power=20 dBm
          Retry  long limit:7   RTS thr:off    Fragment thr:off
          Encryption key:off
          Power Management:off

eth0      no wireless extensions.

root@bt:~#
```

Finally we have a wireless card that is in BackTrack and is recognized! This, as you have discovered, is no small feat. Let us now review some of the specifications of what we are examining here. The device is **wlan0**, it supports 802.11 b/g, and as of now, it is not associated with any **ESSID (Extended Service Set Identification)**. The mode that the card is in is called **Managed**, which we will discuss later. The power is set at **20 dBm**. For most of you, this is probably not how you understand power measurement; this is because most of us think of power in terms of milliwatts (mW) when it comes to wireless devices. For those of you who would like to know the math, dBm is decibel-milliwatt, and is the electrical power unit in decibels (dB) referenced to 1 milliwatt (mW). The formula is as follows:

- The power in decibel-milliwatts is equal to the base 10 logarithm of the power in milliwatts (P(mW))

  $P(dBm) = 10 \log 10(P(mW) / 1mW)$

- The power in milliwatts (P(mW)) is equal to 10 raised by the power in decibel milliwatts, P(dBm), divided by 10

  $P(mW) = 10(P(dBm)/10)$

For most of us, this is way too much to think about. For that, there is a table that we can use as a reference, which is as follows:

Power (dBm)	Power (dBW)	Power (watt)	Power (mW)
20 dBm	-10 dBW	100 mW	100 mW
30 dBm	0 dBW	1 W	1000 mW

From the table, you can see that 20 dBm is equivalent to 100mW, and that is the default setting currently on `wlan0`. We can of course change this, but for now, we will leave it as it is. It is also important to note that the maximum power you can legally transmit is 1 Watt or 1000 mW in the U.S.; if you refer to the table, you can see that it is equivalent to 30 dBm.

Now that we have the particulars out of the way, we can bring up our network card. We do this using the versatile `ifconfig` command. Enter `ifconfig wlan0 up` in the terminal window:

```
root@bt:~# ifconfig wlan0 up
```

The command should complete and not provide any feedback (you will not see any output, just the command prompt window) but the card has started. You can verify this by repeating our `ifconfig` command:

```
root@bt:~# ifconfig
```

The output of the command with the card up and working is shown in the next screenshot:

```
lo          Link encap:Local Loopback
            inet addr:127.0.0.1  Mask:255.0.0.0
            inet6 addr: ::1/128 Scope:Host
            UP LOOPBACK RUNNING  MTU:16436  Metric:1
            RX packets:1287 errors:0 dropped:0 overruns:0 frame:0
            TX packets:1287 errors:0 dropped:0 overruns:0 carrier:0
            collisions:0 txqueuelen:0
            RX bytes:397909 (397.9 KB)  TX bytes:397909 (397.9 KB)

wlan0       Link encap:Ethernet  HWaddr 00:60:b3:05:6d:eb
            UP BROADCAST MULTICAST  MTU:1500  Metric:1
            RX packets:0 errors:0 dropped:0 overruns:0 frame:0
            TX packets:0 errors:0 dropped:0 overruns:0 carrier:0
            collisions:0 txqueuelen:1000
            RX bytes:0 (0.0 B)  TX bytes:0 (0.0 B)
```

As you refer to the previous screenshot, notice that `lo` and `wlan0` represent names of the network interfaces on the machine, with `lo` referring to the loopback address and `wlan0` referring to the wireless card. You can see that `lo` has an INET address. This is because `lo` is configured as `127.0.0.1` on the machine. The `wlan0` card does not have this line because an address has not been assigned.

Wireless card modes

At this point, it is important that we establish what we want to accomplish with the BackTrack tool. We really are not interested in using the wireless card to connect to an access point, because what we want to do is to use the card differently. When you connect a wireless card to an access point, you are using one of the modes of the card. That mode is the managed mode, which is the mode that the card is in, as shown earlier. When you want to analyze network traffic (or "sniff", as it is commonly referred to), you put the card in a promiscuous mode. With our wireless cards, we do not call the mode promiscuous; we instead call it the monitor mode. The best way to put a card into the monitor mode is to use a tool, and we will do that in a moment. First off, we will see how to use the BackTrack card to connect to the network, just in case at times you want to be able to do this.

The tool we will use now is the one that is GUI-based. We could do the entire configuration without using the GUI, but as our main focus is analyzing the traffic, we will stick with the GUI for this *one* time. The tool we are going to use is the Wicd Network Manager. We can access this tool from the desktop menu by going to **Applications | Internet | Wicd Network Manager**. This should open the tool, and you should see a window similar to the one shown in the next screenshot:

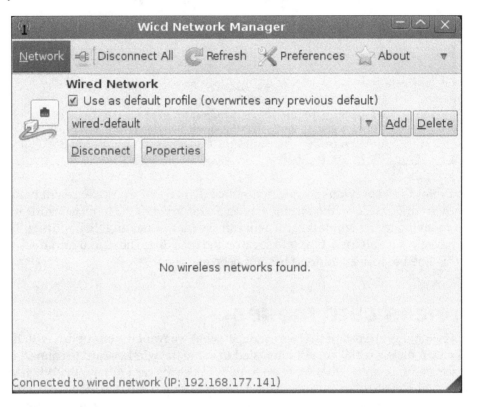

Oh no! We still do not have a wireless network. What are we going to do? Why is this so complicated? Before you panic, let us look at the reason why no wireless networks are found. Remember, we knew that our wireless card was on because we verified it from the command line. This is the catch with these things; we should verify them on the command line first and then work with the GUI. The way to look at this is, the GUI is not really that intelligent because it is going to reflect whatever the programmer has set up by default. We do not have this problem on the command line because we are controlling the actual lowest level of the device.

With a GUI, this may or may not be the case, depending on what the programmers were thinking when they wrote the code. Coming from a programming background, I can honestly say that most programmers just try to get the code to work and meet the objectives of the specification, and they cannot determine what is in the user's mind. The first thing you see when you look at the window is that it is based on a profile, and that the default profile is wired; so there is a possibility that this profile does not include the wireless information that we need. Let us now look deeper into this. Click on **Preferences**. This will open the window that allows us to set preferences, and it is displayed in the next screenshot:

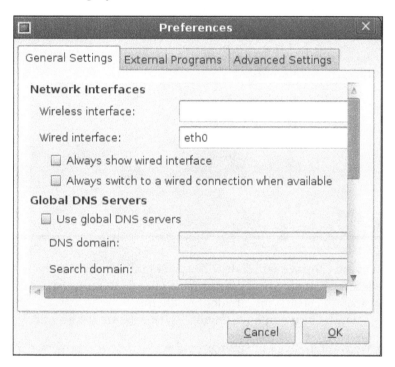

See! All our panic was for nothing, because the configuration in the default state does not have a wireless interface listed. So if we put the wireless interface in, maybe this will work! Unfortunately, "maybe" does not cut it when it comes to wireless in BackTrack, so what we will do is we will verify it! Remember, we verify all assumptions when it comes to security!

Enter wlan0 in the box for **Wireless interface:** as shown in the following screenshot:

It should be obvious, but we will state it here just to make sure; if you have another name for the interface, then you should enter whatever the interface is called instead of entering wlan0. To check, you can identify the name of the interface using the ifconfig command. Once you have entered the necessary information, click on **OK**.

At this point, we want to save our settings, and we will do this now by clicking on the **Add** button. Enter the name wireless for your profile, and click on **OK**. You will notice that we still do not have any wireless networks showing. We need to click on the **Refresh** button, and then the tool should scan and look for our wireless networks. An example of this is shown in the following screenshot:

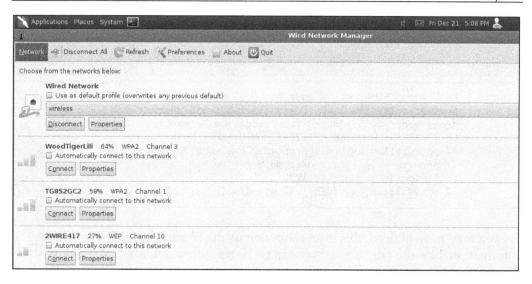

Protocol analysis with the network card

It is disturbing that we continue to see networks using the Wired Equivalent Privacy (WEP) algorithm. We will get into the weaknesses of WEP and show how easy it is to obtain the WEP key for an attacker and also for you if you are working with your own network or have written the authorization to test someone else's network. There is absolutely no reason you or anyone else should be using WEP to protect a wireless network. Now that you know how to get your card to work, you can click on a network and connect to it if you like by clicking on the **Connect** button. An example of a connection to the **WoodTigerLili** network is shown in the following screenshot:

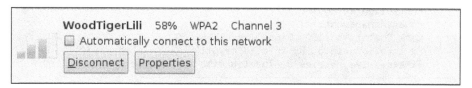

Once we are connected, we can treat this like any other wireless connection. You can verify that it is connected by entering the `ifconfig wlan0` command in a terminal window:

```
root@bt:~# ifconfig wlan0
```

The output of a connected adapter is shown in the next screenshot:

```
                                   root@bt: ~                               _ ^ x
File  Edit  View  Terminal  Help
root@bt:~# ifconfig wlan0
wlan0       Link encap:Ethernet  HWaddr 00:60:b3:05:6d:eb
            inet addr:192.168.1.13  Bcast:192.168.1.255  Mask:255.255.255.0
            inet6 addr: fe80::260:b3ff:fe05:6deb/64 Scope:Link
            UP BROADCAST RUNNING MULTICAST  MTU:1500  Metric:1
            RX packets:100 errors:0 dropped:0 overruns:0 frame:0
            TX packets:26 errors:0 dropped:0 overruns:0 carrier:0
            collisions:0 txqueuelen:1000
            RX bytes:10311 (10.3 KB)  TX bytes:4710 (4.7 KB)
```

As you may have noticed, this is just showing us the IP address and network information, like any other card. Wouldn't it be better to see the actual wireless information? Fortunately, we have a command for this as we do for most of the things we want to do. Enter the following command to see the information on the wireless network:

root@bt:~# iwconfig wlan0

A sample of the output from this command is shown in the next screenshot:

```
                                   root@bt: ~                               _ ^ x
File  Edit  View  Terminal  Help
root@bt:~# iwconfig wlan0
wlan0       IEEE 802.11bg  ESSID:"WoodTigerLili"
            Mode:Managed  Frequency:2.422 GHz  Access Point: E0:91:F5:BB:30:61
            Bit Rate=48 Mb/s    Tx-Power=20 dBm
            Retry  long limit:7    RTS thr:off    Fragment thr:off
            Encryption key:off
            Power Management:off
            Link Quality=56/100  Signal level=56/100
            Rx invalid nwid:0  Rx invalid crypt:0  Rx invalid frag:0
            Tx excessive retries:0  Invalid misc:9   Missed beacon:0
```

Earlier, we discussed that there are a number of modes when it comes to a wireless card. There are only two that are of concern to us, and they are as follows:

- Managed

  The mode for connecting to an access point

- Monitor

  The mode for "sniffing" network data

Once you are in the managed mode on a network card in Linux, you cannot see the wireless network traffic; this is why we use an external card in addition to the card that we have on our machine. The reason for this is we want to be able to analyze our wireless traffic using the powerful tools that are contained within the BackTrack distribution. So, for our purposes, we will not be connecting to our wireless card within BackTrack; we will leave the card in the monitor mode for using our tools. We will now verify that we are only able to see the Ethernet (802.3) data when we are connected, and not the wireless (802.11) traffic, which is what we want to be able to use once we unleash the powerful tools of BackTrack. What this means is with 802.3 traffic, you will see the normal web traffic and other normal network information; with 802.11 traffic, you will see the wireless data that is used to communicate over wireless networks. To view network packets, we will turn to a very powerful protocol analyzer tool that is not only available in BackTrack but also available for Windows and is *free*! The tool we will use is named Wireshark (www.wireshark. org). Wireshark is a powerful protocol analyzer that allows us to look at the network traffic. We can access the tool in BackTrack by going to **Applications | BackTrack | Information Gathering | Network Analysis | Network Traffic Analysis | Wireshark**. An example of this is shown in the next screenshot:

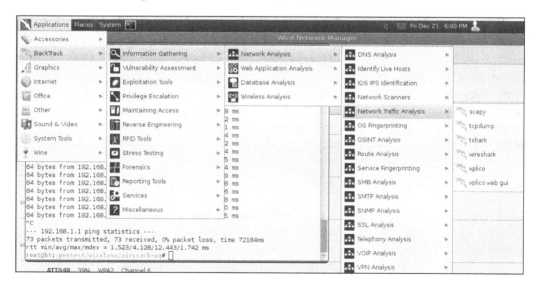

Once you have done this, you will see the application open, after which you will be in the main window. Once you are in the main window, you should see an area on the left-hand side of the screen that provides a listing of the interfaces that are available on the machine. As we have done before, we want to use the interface that is used for our wireless connection, `wlan0`. If you get a window that pops up and displays a warning that you are running as the root user, click on **OK** to acknowledge the message, and then click and place a checkmark in the **Don't show this message again** checkbox. An example of the Wireshark main window is shown in the following screenshot:

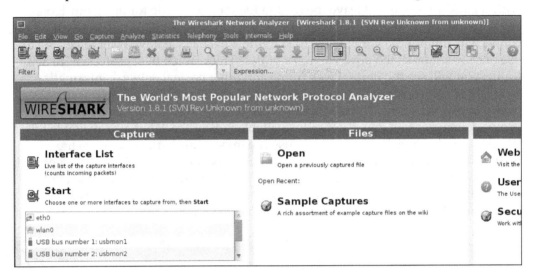

Select the `wlan0` interface, and click on **Start**. You should now be ready to capture some packets. As what you are capturing depends on your settings, you may or may not be seeing traffic; so in either case, it is a simple matter of generating some traffic. In your BackTrack tool, go to **Applications** | **Accessories** | **Terminal** and open a terminal window. When the window opens, enter `ifconfig` and determine what network your `wlan0` interface is on.

```
root@bt:~# ifconfig
```

This will be indicated by the INET address, which is the IP address of your wireless network card. Once you have discovered this, you just want to enter `ping <ip address of the network>`. If you are in a virtual machine, the host machine is usually the first node. The following screenshot shows a ping to the host machine from the BackTrack machine:

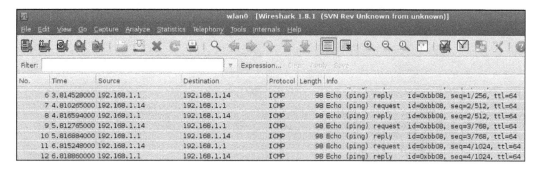

```
root@bt:~# ifconfig wlan0
wlan0     Link encap:Ethernet  HWaddr 00:60:b3:05:6d:eb
          inet addr:192.168.1.14  Bcast:192.168.1.255  Mask:255.255.255.0
          inet6 addr: fe80::260:b3ff:fe05:6deb/64 Scope:Link
          UP BROADCAST RUNNING MULTICAST  MTU:1500  Metric:1
          RX packets:1635 errors:0 dropped:0 overruns:0 frame:0
          TX packets:526 errors:0 dropped:0 overruns:0 carrier:0
          collisions:0 txqueuelen:1000
          RX bytes:167507 (167.5 KB)  TX bytes:99683 (99.6 KB)

root@bt:~# ping 192.168.1.1 -c 3
PING 192.168.1.1 (192.168.1.1) 56(84) bytes of data.
64 bytes from 192.168.1.1: icmp_seq=1 ttl=64 time=4.48 ms
64 bytes from 192.168.1.1: icmp_seq=2 ttl=64 time=3.44 ms
64 bytes from 192.168.1.1: icmp_seq=3 ttl=64 time=4.28 ms
```

We should now be able to see the data that Wireshark captured. Note that if you do not add the -c option to your ping request, it will ping forever — well not really forever — but it will not stop without intervention and you will have to stop it by pressing *Ctrl + C*. In case you are not aware of this key combination, it will get you out of virtually every program that is running in Linux. In your Wireshark application, click on **Stop** under **Capture**. You should see your ICMP packet information as shown in the following screenshot:

Summary

This concludes the chapter. In this chapter, we got to experiment with getting your wireless card working in BackTrack; as you discovered, it is no small task. We also saw the recommendation that you make a small investment and get one of the cards that is supported by BackTrack. We saw two USB cards, Wi-Fire and Alpha. Either of these cards will allow you to use the full power of the BackTrack tools. In the next chapter, we will look at how to use the tools to explore the wireless zones around you.

3
Surveying Your Wireless Zone

In this chapter, we are going to look at the following with respect to using Backtrack:

- Exploring devices
- Using Kismet
- Working with different tools
- Discovering access points

We should now have a working wireless card that will allow us to start using the powerful tools within Backtrack, which will in turn allow us to survey our wireless zones. When we refer to surveying the wireless zones, we mean looking at the strength of the signal and the channel that the access point is using, along with the security that is set on the access point.

We now want to start up our Backtrack machine, log in to it with a username of `root` and a password that you have set; if you are still using the default password, it is `toor`.

Once the Backtrack tool has started, we want to start the window environment; as a quick reminder, enter `startx`:

```
root@bt:~# startx
```

Once the window environment comes up, we want to verify that our network card is recognized in the Backtrack tool. Open a terminal window by going to **Applications | Accessories | Terminal**.

In the terminal window, enter the following command:

```
root@bt:~# iwconfig
```

This should result in the output you see in the next screenshot if your wireless card is active and recognized by Backtrack:

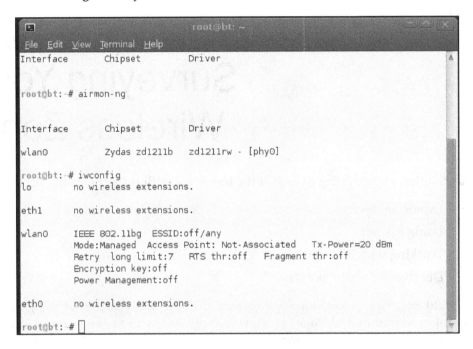

If your card is not recognized, that is, you are not able to see wlan0, refer to the procedures and techniques we used in the previous chapter to get your card working.

Exploring devices

The next thing we want to do is explore the wireless networks that are around us. The first tool we will use within BackTrack is not actually a hacking tool, but it does work for surveying the networks around us and identifying our wireless zone. Navigate to **Applications** | **Internet** | **Wicd Network Manager**.

This will open the tool; if you get a bus error, click on **OK**. The tool should open and present a window similar to the one shown in the next screenshot:

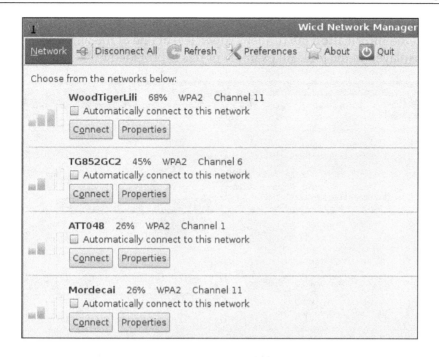

As you can see, the Wicd Network Manager does provide the channel and additional information about the card, so before we go further, we need to discuss what the channels are. In wireless, we have a particular frequency on which the signal is propagated; if everyone is on the same frequency, the result is saturation. So to avoid that, we use different channels for separation and to not overload or saturate a channel. You are probably wondering what channel you should use. Well, this can be a complicated question. This is why we are using this chapter to see how to survey a wireless zone. If everyone is on the same channel near you, it can lead to a degradation in the performance of your wireless network.

A bit about channels — there are a number of channels that are available, which change depending on the version of the 802.11 standard that you are using; so rather than try and explain it here, the Wikipedia site has a good definition on the different channels and which countries use what channels, and you can find that information at http://en.wikipedia.org/wiki/List_of_WLAN_channels.

The next thing we want to do is explore the access point information further, so click on the **Information** tab (under **Properties**) of one of the wireless networks that are displayed in the tool. An example of this is provided in the next screenshot:

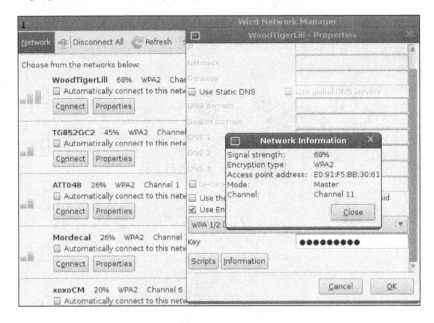

From the screenshot, you can see that the signal strength of the WoodTigerLili access point is at **68%**, it is using the **WPA2** encryption scheme, and it is on **Channel 11**. In fact, in this example, all of the discovered access points are using WPA2, which is a good thing with respect to security.

As you may recall from earlier, we have to put the card in promiscuous mode, and the mode we are currently in is the monitor mode of the card. To do this, we use the `airmon-ng` tool. In a terminal window, enter the following command:

```
root@bt:~# airmon-ng start <network interface>
```

This command will start the monitor mode on the interface that has the wireless card. In this example, it is on `wlan0`. If your card is on another interface, you need to enter that for the command. Once you have entered the command, you should have an output similar to the one shown in the following screenshot:

```
root@bt:/pentest/wireless/aircrack-ng# airmon-ng start wlan0


Found 3 processes that could cause trouble.
If airodump-ng, aireplay-ng or airtun-ng stops working after
a short period of time, you may want to kill (some of) them!

PID      Name
1404     dhclient3
1414     dhclient3
2132     dhclient3
Process with PID 2132 (dhclient3) is running on interface wlan0


Interface      Chipset        Driver

wlan0          Zydas zd1211b  zd1211rw - [phy0]
                              (monitor mode enabled on mon0)
```

From the output of the command, you can see that the card is in monitor mode; this is now reflected as interface mon0. We will, from this point onwards, use the mon0 interface when we want to process data from the card.

The first command-line tool we will look at is the ssidsniff tool. Go to **Applications | Backtrack | Information Gathering | Wireless Analysis | WLAN Analysis | ssidsniff**.

In the window that opens, enter the following command:

```
root@bt:~# ssidsniff -i mon0
```

The -i refers to the interface to which to bind; in this case, it is mon0, which represents the interface name. This should open the tool, and it will scan and display the networks as it sees them. An example of this is shown in the next screenshot:

```
                                      root@bt: ~
File  Edit  View  Terminal  Help
ssidsniff | up 00:12:22  14 networks
mon0: 33.9K pkts: 22.6K beacons, 3.0K data, 2.5K other

SSID Name                  BSSID              Flg  Cn Pkts   Data  PPS  Sig
>ATT408                    90:b1:34:d1:3e:90  AEx  1  7.5K   1.5K  50
ATT048                     20:e5:64:b1:a7:d0  AEx  1  7.9K   672   8
ATT328                     40:b7:f3:46:4c:20  AEx  1  7.8K   692   8
ATT304                     40:b7:f3:cf:d3:10  Awx  1  1.9K         2
Chateau Blues and Brew20:10:7a:16:68:9f       AEx  1  1.3K   57
ATT336                     40:b7:f3:6f:6b:80  AEx  1  512    83
ATT360                     38:6b:bb:00:5d:20  AEx  1  119    1
eli                        20:10:7a:13:d5:b6  AEx  1  54     1
SBG658084                  20:10:7a:13:d1:8b  Awx  1  1
                           00:25:4b:04:bc:89  H       316
ATT864                     90:b1:34:b1:02:70  Awx  1  14
GBCGMEN2.4                 e0:91:f5:ec:21:0b  AEx  1  317    12
belkin.496                 ec:1a:59:06:d4:96  Awx  1  5
                           e0:91:f5:bb:30:61  H       9
```

You can use your arrow keys to go through the list; once you have selected a particular SSID, you can explore further by pressing the *Enter* key. The result of this will be another window that shows additional details about the access point. An example of this is provided in the following screenshot:

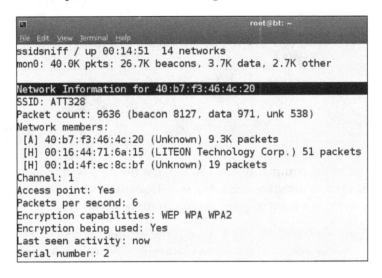

The tool shows you information about the access point and also the machines that are connected to the access point at this time. If you press the *D* key, any strings that are discovered within the packets will be displayed. Of course, this will not show anything if all of the packets are encrypted.

You can sort the access points in a number of ways. By pressing the *O* key, you can select the field on which to sort; for example, press *C* if you want to sort by channel. A list that is channel-sorted is shown in the following screenshot:

```
ssidsniff | up 00:24:41  20 networks
mon0: 2.462GHz, 66.1K pkts: 43.7K beacons, 5.9K data, 4.7K other

SSID Name                    BSSID              Flg  Cn Pkts   Data  PPS Sig
ATT408                       90:b1:34:d1:3e:90  AEx  1  13.5K  2.8K
ATT048                       20:e5:64:b1:a7:d0  AEx  1  14.7K  1.2K
ATT328                       40:b7:f3:46:4c:20  AEx  1  14.3K  1.3K
ATT304                       40:b7:f3:cf:d3:10  Awx  1  3.3K
Chateau Blues and Brew20:10:7a:16:68:9f         AEx  1  1.7K   94
ATT336                       40:b7:f3:6f:6b:80  AEx  1  730    142
ATT360                       38:6b:bb:00:5d:20  AEx  1  213    3
>eli                         20:10:7a:13:d5:b6  AEx  1  57     1
SBG658084                    20:10:7a:13:d1:8b  Awx  1  1
                             00:25:4b:04:bc:89  HE      461
ATT864                       90:b1:34:b1:02:70  Awx  1  16
GBCGMEN2.4                   e0:91:f5:ec:21:0b  AEx  1  432    12
belkin.496                   ec:1a:59:06:d4:96  Awx  1  5
WoodTigerLili                e0:91:f5:bb:30:61  AEx  11 1.2K   41    8
Full Monty                   00:26:50:63:8d:99  AEx  1  4      2
Mordecai                     20:10:7a:12:9d:94  AEx  11 1.2K   267   10
VIGER                        e0:91:f5:df:1d:a1  AEx  11 1.1K   2     9
```

There are quite a few options with which we can experiment in this tool, but we want to look at some more tools. As always, you are encouraged to explore beyond what we have done in the book.

Working with tools

The next tool we want to work with is the `airodump-ng` tool. This is another powerful tool with many features that is available within Backtrack. We will look at this tool in more detail in the next chapter, but for now we will use it within the context of surveying our wireless zone. Go to **Applications | Backtrack | Information Gathering | Wireless Analysis | WLAN Analysis | airodump-ng**.

Once the window opens, enter the following command:

```
root@bt:~# airodump-ng mon0
```

This should result in a window that will display information about the access points that are being detected by the card. An example of this is shown in the next screenshot:

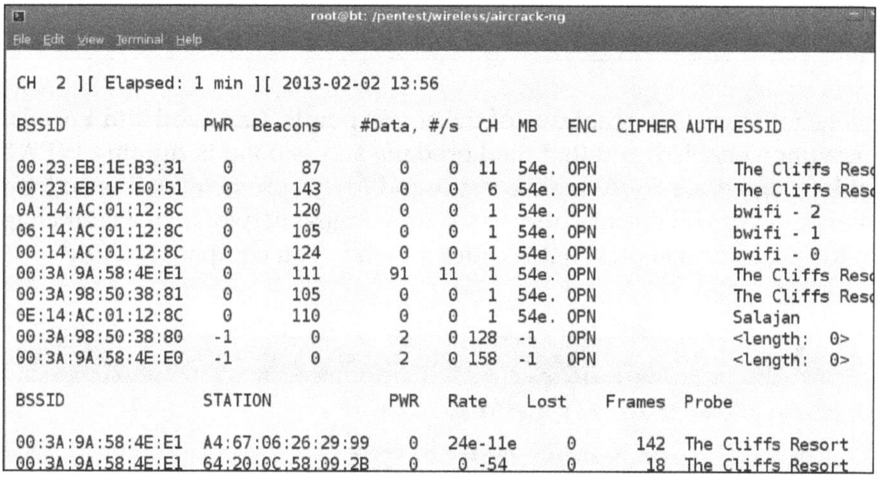

As you can see from the screenshot, the output is similar to that of the `ssidsniff` tool. One of the features that `airodump-ng` has is, it can be used with a Global Positioning System (GPS) receiver; it will then display the coordinates of the discovered access points. In the previous screenshot, all of the access points that have been discovered are open, as shown below the **ENC** field. As such, we do not get to see the encrypted data that is shown by the tool. We will take a look at that now. The next screenshot shows the `airodump-ng` tool with a discovered access point that has encryption enabled.

```
                            root@bt: /pentest/wireless/aircrack-ng                    _  v
File  Edit  View  Terminal  Help

 CH 10 ][ Elapsed: 8 s ][ 2013-02-02 14:14

 BSSID              PWR  Beacons    #Data, #/s  CH  MB   ENC  CIPHER AUTH ESSID

 00:23:EB:1E:B3:31   0        6        0    0   11  54e. OPN              The Cliffs Reso
 00:23:EB:1F:E0:51   0       12        0    0    6  54e. OPN              The Cliffs Reso
 0E:14:AC:01:12:8C   0       12        0    0    1  54e. OPN              Mocanu
 00:3A:98:50:38:81   0       17        0    0    1  54e. OPN              The Cliffs Reso
 0A:14:AC:01:12:8C   0       14        0    0    1  54e. WEP  WEP         Cardwell
 00:3A:9A:58:4E:E1   0       12        1    0    1  54e. OPN              The Cliffs Reso
 06:14:AC:01:12:8C   0       14        0    0    1  54e. OPN              Kevin
 00:14:AC:01:12:8C   0       19        0    0    1  54e. WPA  TKIP   PSK  Loredana

 BSSID              STATION            PWR   Rate    Lost    Frames  Probe

 00:3A:9A:58:4E:E1  A4:67:06:26:29:99   0    0 -11    559       3
```

The screenshot now shows that two of the access points, **Cardwell** and **Loredana**, have encryption enabled, and that the **Loredana** access point is running **WPA TKIP**. Not that long ago there were weaknesses found in this protocol, so it is no longer recommended. We will discuss how to secure wireless networks in a later chapter. Refer to the next screenshot. What is different when you compare it to the previous image?

```
                            root@bt: /pentest/wireless/aircrack-ng                    _  v
File  Edit  View  Terminal  Help

 CH 11 ][ Elapsed: 20 s ][ 2013-02-02 14:24

 BSSID              PWR  Beacons    #Data, #/s  CH  MB   ENC  CIPHER AUTH ESSID

 00:23:EB:1E:B3:31   0       14        0    0   11  54e. OPN              The Cliffs Reso
 00:23:EB:1F:E0:51   0       14        2    0    6  54e. OPN              The Cliffs Reso
 0E:14:AC:01:12:8C   0       20        0    0    1  54e. OPN              Mocanu
 00:3A:98:50:38:81   0       21        0    0    1  54e. OPN              The Cliffs Reso
 0A:14:AC:01:12:8C   0       25        0    0    1  54e. WEP  WEP         Cardwell
 00:3A:9A:58:4E:E1   0       18       46   10    1  54e. OPN              The Cliffs Reso
 06:14:AC:01:12:8C   0       25        0    0    1  54e. OPN              <length:  5>
 00:14:AC:01:12:8C   0       23        0    0    1  54e. WPA  TKIP   PSK  Loredana

 BSSID              STATION            PWR   Rate    Lost    Frames  Probe

 00:23:EB:1F:E0:51  F0:A2:25:F4:5E:3B  -1   18e- 0      0       2
 00:3A:9A:58:4E:E1  24:77:03:07:46:F0   0    0 - 1e    22       3
 00:3A:9A:58:4E:E1  00:1B:63:CD:FC:C1   0    0 -54e    13      25
 00:3A:9A:58:4E:E1  A4:67:06:26:29:99   0   48e-48      0      24
```

Have you figured it out? If you look at the penultimate access point, you will notice that there is no SSID name there; however, you do see the tool has detected that it is a five-character SSID, but it does not know what it is. This access point has been configured to cloak its SSID. This is a setting that many will use to hide their access point, but it is not that effective, as we will now show. If you look at the next screenshot, can you determine the name of that access point?

```
File  Edit  View  Terminal  Help

CH  1 ][ Elapsed: 40 s ][ 2013-02-02 14:29

BSSID              PWR  Beacons    #Data, #/s  CH  MB   ENC  CIPHER AUTH ESSID

00:23:EB:1F:E0:51    0       30        4    0   6  54e. OPN              The Cliffs Reso
00:23:EB:1E:B3:31    0       54        0    0  11  54e. OPN              The Cliffs Reso
06:14:AC:01:12:8C    0       41       25    0   1  54e. OPN              <length:  5>
00:14:AC:01:12:8C    0       47       11    0   1  54e. WPA  TKIP   PSK  Loredana
0E:14:AC:01:12:8C    0       53       16    0   1  54e. OPN              Mocanu
00:3A:98:50:38:81    0       42        0    0   1  54e. OPN              The Cliffs Reso
00:3A:9A:58:4E:E1    0       42        0    0   1  54e. OPN              The Cliffs Reso
0A:14:AC:01:12:8C    0       47       16    0   1  54e. WEP  WEP         Cardwell
00:3A:9A:58:4E:E0   -1        0        2    0 113  -1   OPN              <length:  0>

BSSID              STATION             PWR   Rate    Lost    Frames  Probe

00:23:EB:1F:E0:51  F0:A2:25:F4:5E:3B   -1   12e- 0       0      4
06:14:AC:01:12:8C  64:80:99:34:02:68    0    0 - 1e     92     36   Kevin
00:3A:9A:58:4E:E1  A4:67:06:26:29:99    0    0 -11       0     16   The Cliffs Resort
```

As you can see in the previous screenshot, in the **Probe** portion of the packet, when the access point is connected to by a client, the SSID is in the response packet. The only protection a cloaked or disabled broadcast of the SSID provides you is if no one connects; because once they connect, the SSID is revealed. In this case, the client station with the MAC address of **64:80:99:34:02:68** has connected and provided us with the missing SSID of the access point named **Kevin**.

Using Kismet for access point discovery

The next tool we will look at is Kismet. This tool will automatically provide us with the SSID when someone connects to a cloaked network. Go to **Applications** | **Backtrack** | **Information Gathering** | **Wireless Analysis** | **WLAN Analysis** | **Kismet**.

You will get an error, because you first need to start the Kismet server. So what we want to do is start kismet_server. You will get a prompt to start the server; click on **Yes**. If the prompt does not come up, you can start the server on your own by entering the following command in a terminal window:

```
root@bt:~# kismet_server -s
```

You will get a number of errors, mainly because of the GPS; for our purposes, we will ignore them. The next thing you want to do, if you started the server from the terminal window, is quit the client. If you started the server within the GUI, you can skip the next step and remain in the client window. To quit the client, click on **Close window** under the **File** menu. Once it has exited, you want to open the tool again by navigating to **Applications | Backtrack | Information Gathering | Wireless Analysis | WLAN Analysis | Kismet**. If you did not start the server from the command line, you can remain in the client tool. This should show that you are connected to the server now. An example of this is shown in the following screenshot:

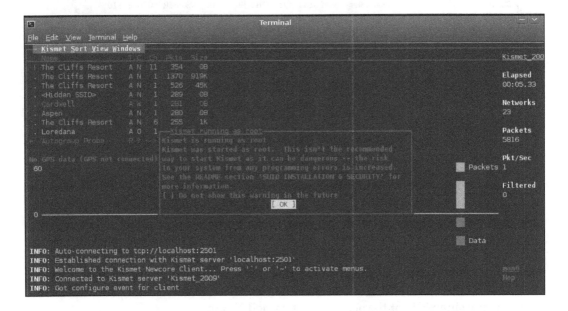

As you can see from the warning, Kismet is running as root, and we really never want to do that; but for now, we will go on, because we are just working with the tool. If you do not see any networks, you will have to specify the mon0 interface. Click on **OK** to close the warning about running as root. If prompted to add the source, you can enter mon0, and then click on **OK**. Alternatively, you can go to **Kismet | Add Source** and then enter the interface mon0.

As you can see in the next screenshot, there is a hidden SSID. Unlike our airodump-ng tool, the Kismet client does not tell us the number of characters for the SSID.

The Kismet tool can be used for many things; it is an excellent packet-capturing tool with a very good logging feature. For a good description and discussion on the tool to identify whether a policy is being adhered to, refer to the following link:

```
http://scholarworks.sjsu.edu/cgi/viewcontent.
cgi?article=1234&context=etd_projects
```

We can use the Kismet log to identify the network ranges of the access points, as well as for a variety of additional information. An example of this is shown in the following screenshot:

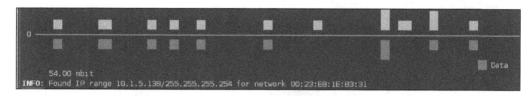

Click on one of the access points that are listed, and then press the *Enter* key; this will bring up a window with the detailed access point information. The example in the next screenshot reflects detailed information about the **Loredana** access point:

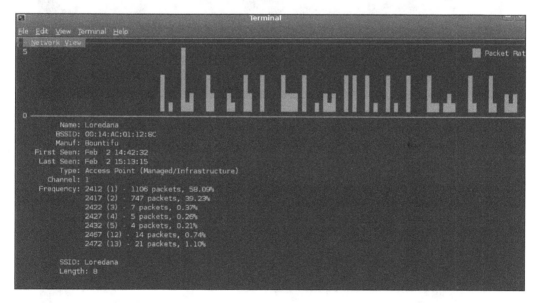

We can also view the clients that are associated with the networks. To do this, click on **Clients** under the **View** menu; this will display the clients that are connected to the network. The menu selection for this is reflected in the following screenshot:

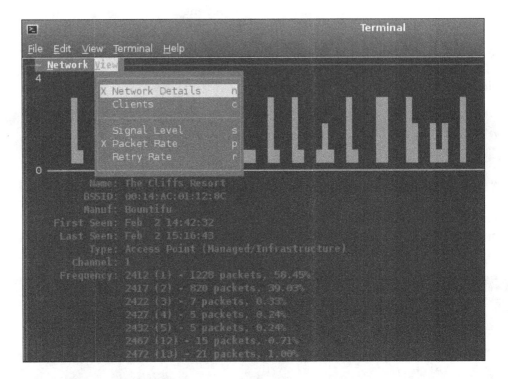

If you select the **Clients** menu option and do not see any clients, you can go to **Network | Next network** and try to discover a network with connected clients. An example of a network that has connected clients is shown in the following screenshot:

On this network, you can see that there are two connected access points; one is from Cisco and the other is from Intel, and they are connected through a wired connection. There are two devices connected; they are from Samsung and Apple. With Backtrack being a penetration testing tool, we could do a lot more with this information. But as we do not have written authorization, we will not do that. The example shown in the next screenshot is of a network for which the SSID is unknown at this time but which has a number of wireless clients attached to it, so this is probably a group of phones connected. Without further investigation, we cannot determine this for sure.

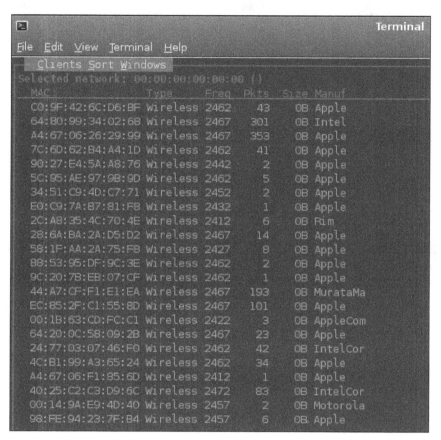

Discovering hidden access points

Now that we have looked at quite a bit of information from the Kismet tool, we want to focus on the feature of discovering access points that are hidden. The way to do this is just like we discussed with the airodump-ng tool. We have to have a client connect to the access point while we are observing it. We have done this, and the Kismet tool output is shown as an example in the next screenshot:

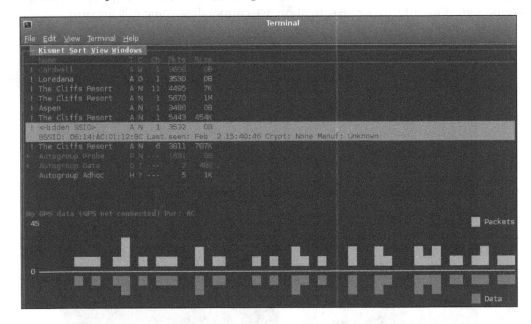

If we press *Enter* on the highlighted SSID, we gain additional information as shown in the following screenshot:

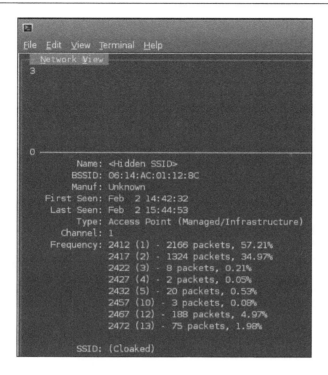

We can also select the connected client to see additional information about the access point. An example of this is provided in the following screenshot:

As you can see from the two preceding screenshots, the SSID is cloaked in the first screenshot, and in the second screenshot, the connected client is the actual access point itself. The next thing we want to do is see what happens when a client connects to the cloaked access point. The response from an access point once a client connects to it, even when it is cloaked, is to include the SSID within the packet. The example that is shown in the next screenshot shows what the Kismet tool discovers when the access point is connected to by clients:

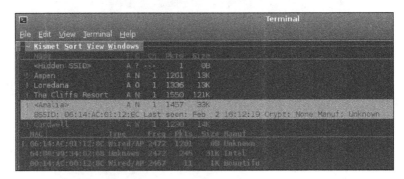

We have now shown how to discover any SSID that is hidden; all that is required is for someone to connect to the access point.

We will look at the Wireshark tool next; as you will remember from earlier, we looked at the Wireshark tool. You can start the Wireshark tool in BackTrack by opening a terminal window and entering `wireshark`:

root@bt:~# wireshark

Once the tool starts, click on the mon0 interface, and then click on the **Start** button to start a packet capture. An example of a beacon that is sent out when the SSID broadcast is disabled is shown in the next screenshot:

```
▷ Frame 202: 120 bytes on wire (960 bits), 120 bytes captured (960 bits) on interface 0
▷ Radiotap Header v0, Length 18
▷ IEEE 802.11 Beacon frame, Flags: ........C
▽ IEEE 802.11 wireless LAN management frame
  ▷ Fixed parameters (12 bytes)
  ▽ Tagged parameters (62 bytes)
    ▷ Tag: SSID parameter set:
    ▷ Tag: Supported Rates 1(B), 2(B), 5.5(B), 11(B), 6(B), 12(B), 24(B), 36, [Mbit/sec]
```

In the previous screenshot, you can see that the SSID parameter set is empty; this is because the SSID broadcast has been disabled. Remember, once a client connects, the response will contain the SSID. An example of this is shown in the next screenshot:

```
◁
Frame 5856: 98 bytes on wire (784 bits), 98 bytes captured (784 bits) on interface 0
▷ Radiotap Header v0, Length 18
▷ IEEE 802.11 Probe Request, Flags: ........C
▽ IEEE 802.11 wireless LAN management frame
   ▽ Tagged parameters (52 bytes)
      ▷ Tag: SSID parameter set: Amalia
      ▷ Tag: Supported Rates 1, 2, 5.5, 11, 6, 9, 12, 18, [Mbit/sec]
      ▷ Tag: Extended Supported Rates 24, 36, 48, 54, [Mbit/sec]
      ▷ Tag: HT Capabilities (802.11n D1.10)
```

This shows that disabling the SSID is a very limited security feature, and as such, is not recommended. Most vendors do not recommend it because it makes it harder to connect to the access points in some cases.

Summary

In this chapter, we looked at a number of tools that we can be used to survey our wireless zone. We have seen how we could detect the channel on which the access points were running as well as the protection scheme that is enabled, be it WEP, or WPA. Additionally, we can also discover which clients are connected to the network, and the IP address range of the network.

Finally, we saw that disabling the SSID broadcast on wireless in itself has limited value, and as such, it is better to leave it enabled.

In the next chapter, we will look at examples of how attackers can break wireless security. It is very important that we discuss how broken security is breached, and then we can see how to protect it.

Breaching Wireless Security

4

In this chapter, we will discuss how hackers attack and breach wireless security. We do this to understand the various methods of attacking so that we can defend against them. There are a number of wireless attacks to discuss on wireless. An important point to understand is that attackers will follow a process and methodology when they attempt an attack, and the way to defend against this is to stop information from getting to the potential attacker. This can be a challenge when it comes to wireless. We group the attacks into several different categories. They are listed in the following table:

Probing and network discovery
Surveillance
Denial of Service
Masquerade
Rogue access point

Different types of attacks

We will now discuss each one of these attacks briefly. The probing and discovery attacks are accomplished by sending out probes and looking for the wireless networks. We have used several tools for discovery so far, but they have all been passive in how they discover information. A passive probing tool can detect the SSID of a network even when it is cloaked, as we have shown with the Kismet tool. With active probing, we are sending out probes with the SSID in it. This type of probing will not discover a hidden or cloaked SSID. An active probing tool for this is **NetStumbler** (www. netstumbler.com). With an active probe, the tool will actively send out probes and elicit responses from the access points to gather information. It is very difficult to prevent an attacker from gathering information about our wireless access points; this is because an access point has to be available for connection. We can cloak or hide the SSID; however, as we have seen in the previous chapter, this is of limited value.

The next step an attacker will carry out is performing the surveillance of the network. This is the technique we used with Kismet, `airodump-ng`, and `ssidsniff`. An example of the output of the Kismet tool is shown in the next screenshot:

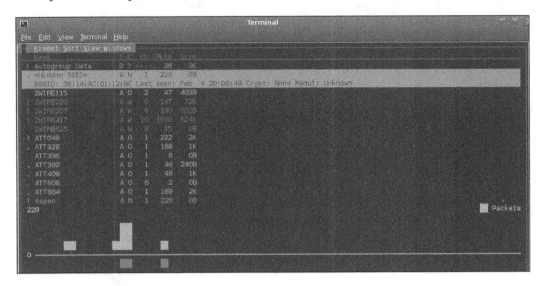

All three of these tools are passive, so they do not probe the network for information. They just capture it from the wireless frequency that is received from the network. Each of these tools can discover the hidden SSID of a network, and again, are passive tools.

Once the attacker has discovered the target network, they will move to the surveillance step and attempt to gather more information about the target. For this, we can again use any of the three tools we previously mentioned. The information that an attacker is looking for are as follows:

- Whether or not the network is protected
- The encryption level used
- The signal strength and the GPS coordinates

When an attacker is scanning a network, he or she is looking for an "easy" target. This is the motive of most of the attackers; they want an easy way in, and almost always, they target the weakest link.

The next step that an attacker will typically pursue is Denial of Service (DoS); unfortunately, this is one area we really cannot do much about. This is because, in the case of a wireless signal, the network can be jammed by using simple and inexpensive tools; so if an attacker wants to perform a DoS attack, there is really not much that we can do to prevent it. So we will not spend any more time on this attack.

The next attack method is one that is shared between the "wired" network world and the wireless world. The attack of masquerading, or spoofing as it is sometimes referred to, involves impersonating an authorized client on a network. One of the protection mechanisms we have within our wireless networks is the capability to restrict or filter a client based on their Media Access Control (MAC) address. This address is that of the network card itself; it is how data is delivered on our networks. There are a number of ways to change the MAC address; we have tools, and we can also change it from the command line in Linux. The simplest way to change our MAC address is to use the `macchanger` tool. An example of how to use this tool to change an address is shown in the next screenshot:

```
root@bt:~# macchanger --mac 00:11:22:33:44:55 wlan0
Current MAC: 00:60:b3:05:6d:eb [wireless] (SMC SMC2642W)
Faked MAC:   00:11:22:33:44:55 (Cimsys Inc)
root@bt:~# airmon-ng start wlan0


Found 3 processes that could cause trouble.
If airodump-ng, aireplay-ng or airtun-ng stops working after
a short period of time, you may want to kill (some of) them!

PID     Name
1404    dhclient3
1414    dhclient3
18979   dhclient3
Process with PID 18979 (dhclient3) is running on interface wlan0


Interface       Chipset         Driver

wlan0           Zydas zd1211b   zd1211rw - [phy2]SIOCSIFFLAGS: Connection
timed out

                                (monitor mode enabled on mon0)


root@bt:~# ifconfig wlan0
wlan0     Link encap:Ethernet  HWaddr 00:11:22:33:44:55
          BROADCAST MULTICAST  MTU:1500  Metric:1
          RX packets:0 errors:0 dropped:0 overruns:0 frame:0
          TX packets:0 errors:0 dropped:0 overruns:0 carrier:0
          collisions:0 txqueuelen:1000
          RX bytes:0 (0.0 B)  TX bytes:0 (0.0 B)

root@bt:~#
```

In the Windows world, we can do it in another way; but it involves editing the registry, which might be too difficult for some of you. The hardware address is in the registry; you can find it by searching for the term `wireless` within the registry. An example of this registry entry is shown in the following screenshot:

0008	abcap	REG_SZ	1
0009	Afterburner	REG_SZ	0
0010	antdiv	REG_SZ	-1
0011	ApCompatMode	REG_SZ	0
0012	AssocRoamPref	REG_SZ	1
0013	band	REG_SZ	0
0014	BandPref	REG_SZ	0
0015	BandwidthCap	REG_SZ	2
0016	BTC_stuck_war	REG_SZ	1
0017	BTCoexist	REG_SZ	1
0018	BusType	REG_SZ	5
0019	ccx_rm	REG_SZ	1
0020	ccx_rm_limit	REG_SZ	300
0021	Chanspec	REG_SZ	11
0022	Characteristics	REG_DWORD	0x00000084 (132)
0023	CoInstallers32	REG_MULTI_SZ	bcmwlcoi.dll, BCMWlanCoInstall
0024	ComponentId	REG_SZ	pci\ven_14e4&dev_4328&subsys_000a1028
0025	Country	REG_SZ	US
0026	DriverDate	REG_SZ	9-20-2007
0027	DriverDateData	REG_BINARY	00 80 ea 2f 19 fb c7 01
0028	DriverDesc	REG_SZ	Dell Wireless 1505 Draft 802.11n WLAN Mini-Card
0029			

The last category of attacks that we will cover here is the rogue access point. This is an attack that takes advantage of the fact that all wireless networks have a particular level of power that they transmit. What we do for this attack is create an access point with more power than the access point we are masquerading as; this results in a stronger signal being received by the client software. When would anyone take a three-bar signal over a five-bar signal? The answer for that would be: never; that is why the attack is so powerful. An attacker can create an access point as a rogue access point; there is no way for most clients to tell whether the access point is real or not. There really is nothing that you can do to stop this attack effectively. This is why it is a common attack used in areas that have a public hotspot. We do have a recommended mechanism you can use to help mitigate the impact of this type of attack, and we will cover that in the next chapter. If you look at the example that is shown in the next screenshot, can you identify which one of the access points with the same name is the correct one?

This is an example of what most clients see when they are using Windows. From this list, there is no way of knowing which one of the access points is the real one.

Now that we have covered, albeit briefly, the steps that an attacker typically uses when preparing for an attack, next we want to look at how to protect the data in transit on a wireless network.

Cracking WEP and WPA

There are a number of ways in which we can protect the data that is transmitted across wireless frequencies. When the IEEE committee formed and created the wireless 802.11 standard, they were conscious enough to know they had to take the security of wireless information into account, because wireless is in the air! Unfortunately, the committee was full of radio engineers, and as such, they really did not know about security. As a result, they made a fatal mistake when they selected the first encryption protocol to protect wireless networks. The flaws of that first protocol, Wired Equivalent Privacy (WEP), are well documented. For more information and an analysis on this, refer to `http://en.wikipedia.org/wiki/Wired_Equivalent_Privacy`.

In short, the selected protocol used an encryption algorithm named **RC4**. It uses a stream cipher, and as such, the key should not be reused. Even the author of the algorithm states this in the paper on the algorithm. Radio engineers should not have been selecting cryptographic algorithms, and if even they were, they should not have done it without some form of peer review. All of these could have prevented the WEP standard from being adopted, but as is often the case when it comes to the IT industry, they were ignored, and we were left with an extremely weak encryption method that provided little security to us. The weaknesses of WEP and how it is broken could fill an entire book, and there are numerous books on this. For this book, we will focus on the attacks not only against WEP, but also WPA.

To crack WEP, we have to capture enough packets with a sufficient number of weak Initialization Vectors (IVs) to perform a number of heuristics and predict the key. The first thing we have to do is discover a WEP-configured access point. We have discussed a number of tools for doing this, and we will not cover them again here. We will concentrate on using the `airodump-ng` tool as it is part of the `aircrack-ng` suite that can be used in conjunction with `airodump-ng` to crack WEP real time. An example of using the `airodump-ng mon0` command is shown in the following screenshot:

```
CH  9 ][ Elapsed: 1 min ][ 2013-02-04 22:10

BSSID              PWR  Beacons   #Data, #/s  CH  MB   ENC  CIPHER AUTH ESSID

20:10:7A:12:9D:94   0        2        3    0  11  54e  WPA2 CCMP   PSK  Mordecai
E0:91:F5:DF:1D:A1   0       12        0    0  11  54e  WEP  WEP          VIGER
00:21:E9:B7:D9:AA   0       14        0    0   5  54e. WPA2 CCMP   PSK  lelan
00:14:95:4C:7A:E2   0       55        0    0   6  54 . WEP  WEP          2WIRE228
00:25:9C:AB:BB:71   0       51        0    0   6  54e. WPA2 CCMP   PSK  MTB
00:14:95:32:24:42   0       56        0    0   6  54 . WEP  WEP          2WIRE287
00:22:75:9E:40:74   0       54        0    0   6  54e  WPA2 CCMP   PSK  xoxoCM
E0:91:F5:BB:30:61   0      109       16    0  11  54e. WPA2 CCMP   PSK  WoodTigerLili
00:1A:C4:4E:3F:D1   0       33        4    0  10  54 . WEP  WEP          Missy1
00:24:56:CE:6C:41   0       82     1020    0  10  11 . WEP  WEP          2WIRE417
00:1D:CF:B3:47:C0   0      118        0    0   6  54e  WPA2 CCMP   PSK  TG852GC2
06:14:AC:01:12:8C   0      113        0    0   6  54e. OPN            Amalia
00:14:AC:01:12:8C   0      101        0    0   6  54e. WPA  TKIP   PSK  TG852GC2
0E:14:AC:01:12:8C   0      117        0    0   6  54e. OPN            Kevin
0A:14:AC:01:12:8C   0      114        0    0   6  54e. WEP  WEP          Cardwell
20:10:7A:13:ED:56   0       92        3    0   6  54e  WPA2 CCMP   PSK  Castelli
90:B1:34:B1:02:70   0        8        3    0   1  54e  WPA2 CCMP   PSK  ATT864
20:10:7A:16:68:9F   0       34        0    0   1  54e  WPA2 CCMP   PSK  Chateau Blues and Brews
90:B1:34:D1:3E:90   0       19       12    0   1  54e  WPA2 CCMP   PSK  ATT408
40:B7:F3:46:4C:20   0       54        3    0   1  54e  WPA2 CCMP   PSK  ATT328
EC:1A:59:06:D4:96   0       18        1    0   1  54e  WPA2 CCMP   PSK  belkin.496
20:10:7A:13:D1:8B   0       26        1    0   1  54e  WPA2 CCMP   PSK  SBG658084
20:E5:64:B1:A7:D0   0       88       53    0   1  54e  WPA2 CCMP   PSK  ATT048
```

As we can see, there are a number of WEP-configured access points, and this is something that we really should not see today. The WEP protocol has been broken since its inception and should not be used on today's networks; this is why the access point manufacturers implement stronger encryption schemes, such as WPA and WPA2.

An attacker could select one of the WEP networks and conduct an attack. Because we are not attackers, we will only attack networks that we have set up for demonstration. From the list in the previous screenshot, we have set up the access point with an ESSID of **Cardwell**.

Within Backtrack, there are a number of tools that we can use to attempt to crack the WEP key. The probability of cracking the key is improved when we have clients connected to the access point. We can still crack the key without connected clients, but the probability is not as high as it is when we have connected clients. The first tool we will use is the **Gerix Wifi Cracker**. You can access the tool by going to **Applications | Backtrack | Exploitation Tools | Wireless Exploitation Tools | WLAN Exploitation | gerix-wifi-cracker-ng**. Once the window opens, the first thing we want to do is enable the monitor mode by navigating to **Configuration | Enable monitor mode**. We then select the mon0 interface and scan for networks. An example of this is shown in the next screenshot:

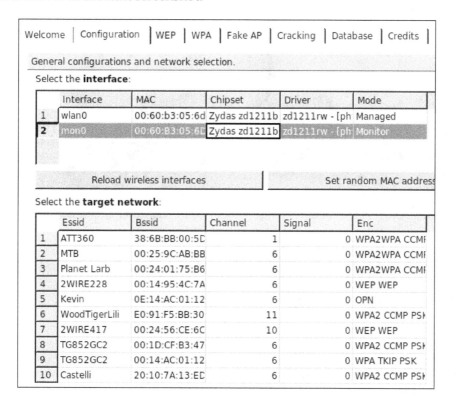

The next thing we want to do is select the WEP network that we want to attack. Once we have selected the network, we will go to **WEP | Start sniffing and logging**. This will open the window that is shown in the next screenshot:

```
                                     sniff_dump --bssid 0A:14:AC:01:12:8C mon0; read;

CH  6 ][ Elapsed: 16 s ][ 2013-02-04 23:32

BSSID              PWR RXQ  Beacons    #Data, #/s  CH  MB    ENC  CIPHER AUTH ESSID

0A:14:AC:01:12:8C   0  25      173        0    0   6  54e.  WEP  WEP         Cardwell

BSSID              STATION            PWR   Rate   Lost    Frames  Probe
```

As the screenshot indicates, the access point **Cardwell** has been selected for our attack. Now that we have started the sniffing and logging, we next want to select our attack, but before we do that, we want to verify that our access point is open for injection. We do this by clicking on the **Performs a test of injection AP**. This should result in the window shown in the next screenshot, which verifies that our injection is possible:

```
    bash -c "aireplay-ng -9 -a 0A:14:AC:01:12:8C mor    _  ^   X
23:39:29  Waiting for beacon frame (BSSID: 0A:14:AC:01:12:8C) on channel 6
23:39:29  Trying broadcast probe requests...
23:39:29  Injection is working!
23:39:31  Found 1 AP

23:39:31  Trying directed probe requests...
23:39:31  0A:14:AC:01:12:8C - channel: 6 - 'Cardwell'
23:39:31  Ping (min/avg/max): 1.414ms/7.579ms/16.268ms Power: 0.00
23:39:31  30/30: 100%
```

Performing an attack using WEP and ChopChop

We are now ready to perform our attacks. We have a number of attacks to choose from. You are probably wondering which one to use? Well, each one has its strengths and weaknesses, and it is mostly a trial and error operation. The process also depends on whether there are clients connected or not. For this demonstration, we will use WEP attacks without clients. We will start with the ChopChop attack. There is a sequence of steps next; they are as follows:

1. In the **ChopChop Attack** section under the **WEP Attacks (no-client)** section, associate with the access point (AP) using fake authentication by clicking on that option (as shown in the following screenshot):

2. Once you have associated with the access point, refer to the **sniff_dump** window. It should look similar to the one in the following screenshot:

3. Click on the **Start the ChopChop attack** option. This will open up another window, which is reflected in the next screenshot:

4. Accept the discovered packet by pressing *Y*. The result of this will be a calculation by the program to generate a file. This is shown in the following screenshot:

```
bash -c "aireplay-ng -4 -h 00:C0:CA:66:7D:CC mo

    Size: 68, FromDS: 1, ToDS: 0 (WEP)

        BSSID  =  00:24:56:CE:6C:41
     Dest. MAC =  FF:FF:FF:FF:FF:FF
     Source MAC =  00:24:56:CE:6C:41

    0x0000:  0842 0000 ffff ffff ffff 0024 56ce 6c41  .B.........$V.lA
    0x0010:  0024 56ce 6c41 8028 05fa ad00 c583 0d05  .$V.lA.(........
    0x0020:  6897 27b7 272c 0469 2209 d6e3 283b 8ab8  h.'.'..i"...(;..
    0x0030:  0978 f6b2 750d 0dff 1d0c fa73 22ef 616c  .x..u......s".al
    0x0040:  3f5b 2e1d                                 ?[..

Use this packet ? y

Saving chosen packet in replay_src-0412-193140.cap

Offset   67 ( 0% done) | xor = 0B | pt = 16 |  431 frames written in  7341ms
Offset   66 ( 2% done) | xor = B0 | pt = 9E |  986 frames written in 16757ms
Offset   65 ( 5% done) | xor = 5A | pt = 01 |  164 frames written in  2790ms
Offset   64 ( 8% done) | xor = A9 | pt = 96 |  180 frames written in  3058ms
Offset   63 (11% done) | xor = 25 | pt = 49 |  278 frames written in  4728ms
Offset   62 (14% done) | xor = 60 | pt = 01 |  112 frames written in  1902ms
Sent 260 packets, current guess: 02...
```

5. In some cases, the attack may fail. If it does fail, you can try again or try the fragmentation attack. Unfortunately, the cracking is not 100 percent. If all goes well, you should see the file get created as shown in the following screenshot:

```
bash -c "aireplay-ng -4 -h 00:C0:CA:66:7D:CC mo
Offset   50 (50% done) | xor = 36 | pt = C0 |  362 frames written in  6149ms
Offset   49 (52% done) | xor = 39 | pt = 41 |  245 frames written in  4168ms
Offset   48 (55% done) | xor = 65 | pt = 6C |   47 frames written in   801ms
Offset   47 (58% done) | xor = 76 | pt = CE |   63 frames written in  1067ms
Offset   46 (61% done) | xor = DC | pt = 56 |  252 frames written in  4282ms
Offset   45 (64% done) | xor = 1F | pt = 24 |  365 frames written in  6210ms
Offset   44 (67% done) | xor = 28 | pt = 00 |  241 frames written in  4092ms
Offset   43 (70% done) | xor = E2 | pt = 01 |  450 frames written in  7655ms
Offset   42 (73% done) | xor = D6 | pt = 00 |   98 frames written in  1663ms
Offset   41 (76% done) | xor = 0D | pt = 04 |  420 frames written in  7139ms
Offset   40 (79% done) | xor = 24 | pt = 06 |   68 frames written in  1157ms
Offset   39 (82% done) | xor = 69 | pt = 00 |  136 frames written in  2319ms
Offset   38 (85% done) | xor = 0C | pt = 08 |  485 frames written in  8243ms
Offset   37 (88% done) | xor = 2D | pt = 01 |  489 frames written in  8301ms
Offset   36 (91% done) | xor = 27 | pt = 00 |  532 frames written in  9054ms
Offset   35 (94% done) | xor = B1 | pt = 06 |  229 frames written in  3891ms
Offset   34 (97% done) | xor = 2F | pt = 08 | 1175 frames written in 19977ms

Saving plaintext in replay_dec-0412-193255.cap
Saving keystream in replay_dec-0412-193255.xor

Completed in 73s (0.41 bytes/s)
```

6. The next step is to create the ARP packet that is to be injected. Click on the **Create the ARP packet to be injected on the victim** access point. Then click on the **Inject the created packet on the victim** access point. This will open another window; accept the packet by pressing *Y*.

7. Now, refer back to the **sniff_dump** window; you should see the data packets counting up. This is reflected in the following screenshot:

```
 □  sniff_dump --bssid 00:24:56:CE:6C:41 mon0; reac  ─ ^ X

CH 10 ][ Elapsed: 14 mins ][ 2013-04-12 19:45

BSSID             PWR RXQ  Beacons    #Data, #/s  CH  MB   ENC  CIPHER AUTH E

00:24:56:CE:6C:41  -56  77    6347      7721  105  10  11 . WEP  WEP       OPN  2

BSSID             STATION          PWR   Rate   Lost    Frames  Probe

00:24:56:CE:6C:41  00:C0:CA:66:7D:CC   0   11 - 1    0     43256
```

The number of data packets (**#Data**) is the important field; if it is counting up, you can start attempting to crack the key when it goes above 5000. You can do this by going to **Cracking | Aircrack-ng – Decrypt WEP password**.

If all goes well, you will see that the key is found, as indicated in the following screenshot:

```
 □           aircrack-log.txt; read; "              ─ ^ X
Opening /root/.gerix-wifi-cracker/replay_dec-0412-193255.cap
Opening /root/.gerix-wifi-cracker/replay_src-0412-193140.cap
Opening /root/.gerix-wifi-cracker/replay_src-0412-194037.cap
Opening /root/.gerix-wifi-cracker/sniff_dump-01.cap
Attack will be restarted every 5000 captured ivs.
Starting PTW attack with 29439 ivs.
                  KEY FOUND! [ 87:27:71:66:43 ]
        Decrypted correctly: 100%
```

In the next chapter, we will discuss the recommended security configurations for the wireless networks of today.

The last attack we will perform is against the WPA protocol; it is an attack against the password used in the configuration of an access point. The method requires the attacker to capture the four-way handshake between a client and an access point, and then load that capture file into a cracking tool to determine the password that was used. One method of doing this is to use the Wireshark program to capture the data and save it to a file. The tool we will use for this is "reaver". We can access the tool by navigating to **Applications | Backtrack | Exploitation Tools | Wireless Exploitation Tools | WLAN Exploitation | reaver**.

Once the window opens, enter the following command:

```
root@bt:~# reaver -i mon0 -b [BSSID] -vv
```

As a reminder, you can get the BSSID from the `airodump-ng` display. After that, it is just a matter of waiting.

The dictionary attack that is used against WPA can be carried out with relative ease once you have captured the handshake. If you know the password, you can verify if it is in the dictionary file, and then run the command to crack it; if it is not there, you have to add it. The tool we will use for this is `aircrack-ng`. As an example, to crack the WPA key, you will load the captured file into the tool as follows:

```
root@bt:~# aircrack-ng -a 2 wpa-psk-linksys.cap -w password.lst
```

This command will try to crack the WPA key based on the defined password list. This is shown in the next screenshot:

```
 File  Edit  View  Terminal  Help

                          Aircrack-ng 1.1 r2178


                  [00:00:00] 233 keys tested (812.16 k/s)


                        KEY FOUND! [ dictionary ]


        Master Key     : 5D F9 20 B5 48 1E D7 05 38 DD 5F D0 24 23 D7 E2
                         52 22 05 FE EE BB 97 4C AD 08 A5 2B 56 13 ED E2

        Transient Key  : 1B 7B 26 96 03 F0 6C 6C D4 03 AA F6 AC E2 81 FC
                         55 15 9A AF BB 3B 5A A8 69 05 13 73 5C 1C EC E0
                         A2 15 4A E0 99 6F A9 5B 21 1D A1 8E 85 FD 96 49
                         5F B4 97 85 67 33 87 B9 DA 97 97 AA C7 82 8F 52

        EAPOL HMAC     : 6D 45 F3 53 8E AD 8E CA 55 98 C2 60 EE FE 6F 51
```

As you can see, cracking the keys is a straightforward process once you have captured the required information or enough packets and weak IVs.

Summary

In this chapter, we have reviewed the steps that an attacker would typically use when targeting or looking for a network. We have identified the weaknesses in the encryption options that were originally used when the 802.11 standard was created. Finally, we showed that the replacement for WEP is WPA, and there are attacks against that as well. We also provided an example of the WPA dictionary attack. In the next chapter, we will revisit these flaws in the wireless protocol and discuss how you can take steps to mitigate the risk of these attacks, and in some cases, eliminate the threat completely.

5
Securing Your Wireless Network

In this chapter, we will look at the following information:

- Configuring initial wireless security
- Adjusting transmit power
- Defending from surveillance
- Configuring encryption

Configuring initial wireless security

In the last chapter, we discussed how an attacker can break into a wireless network. Now that we have discussed this process and methodology, we will look at securing our wireless networks from attacks.

The first steps in the methodology that we discussed were probing and network discovery. As was mentioned in the previous chapter, it is difficult to prevent this because we need to have our networks available. Having said that, there are things we can do to limit our "visibility" when it comes to wireless networks.

Unless you need to broadcast your wireless network out to the public, there is no need to transmit your signals at full power. Not all access points will have the option to reduce the transmission power, but it is worth looking into, to see if yours supports it.

The majority of access points will have a web-based interface that you can use to configure information about the network. To access it is a simple matter of opening a browser and entering the IP address. If you do not know the IP address, note that it is usually the address of the default gateway when you view your wireless connection. On a Windows machine, open a command prompt window, and enter `ipconfig /all | more`. This should result in a list of details of your network cards on that machine. An example of this is shown in the following screenshot:

```
Wireless LAN adapter Wireless Network Connection:

   Connection-specific DNS Suffix   . :
   Description . . . . . . . . . . . : Intel(R) Centrino(R) Ad
   Physical Address. . . . . . . . . : 64-80-99-34-02-68
   DHCP Enabled. . . . . . . . . . . : Yes
   Autoconfiguration Enabled . . . . : Yes
   IPv4 Address. . . . . . . . . . . : 192.168.100.10(Preferre
   Subnet Mask . . . . . . . . . . . : 255.255.255.0
   Lease Obtained. . . . . . . . . . : Tuesday,  February  05,  2
   Lease Expires . . . . . . . . . . : Tuesday,  February  05,  2
   Default Gateway . . . . . . . . . : 192.168.100.1
   DHCP Server . . . . . . . . . . . : 192.168.100.1
   NetBIOS over Tcpip. . . . . . . . : Enabled
```

As you can see from the previous screenshot, the wireless access point is located at the location `192.168.100.1` on this network. The next thing to do is to access the web server that is running on the machine and look at the configuration settings and, more importantly, at the options. We'll do this by opening a browser and entering the IP address that we discovered. For this example, we will enter `http://192.168.100.1`.

If all goes well, we will be presented with a login page for the device. This is shown in the following screenshot:

At this point, we need to know the username and password of the device. In many cases this is admin, with a password of admin. If it has been changed, we could be in for a challenge; if all else fails, we can reset it using the button on the device. There are a number of ways to discover what credentials are required to access the device. Some devices will have the information on the device itself. If you are struggling, you can use the Internet to do a search for the default password of the device. A favorite site to use is www.defaultpassword.com. Once you are on the site, you can enter the manufacturer and product's information and look up the default passwords of the device. If you have a user manual available, you can try using that. If all else fails, you can try to hack into the device. Once you have successfully determined the login credentials, you should be provided access to the device's configuration pages. These pages may vary depending on the device, so it is best to consult the user manual. An example of the configuration page of a Bountiful router is shown in the next screenshot:

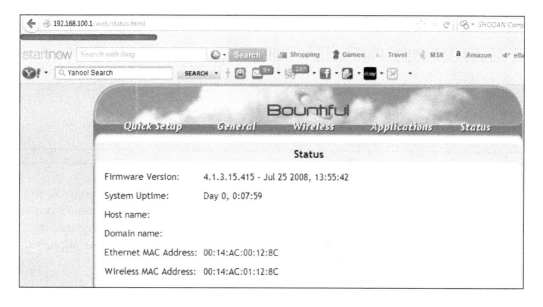

Again, there really is no good way to explore the settings here, so we can either use the user manual, search on the Internet, or just click on things to see what we have. The first thing to click on is usually the tab that has something to do with wireless networks. When you click on the **Wireless** tab of the configuration page for the Bountiful router, you are presented with the screen shown in the next screenshot:

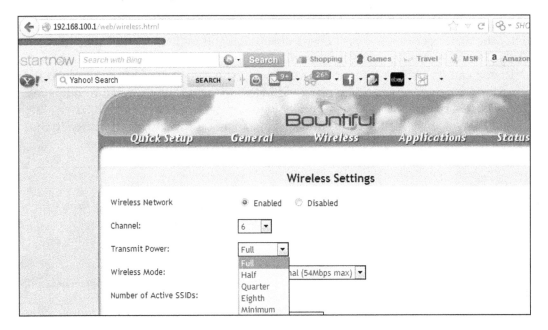

As we can see, the Bountiful router makes it very easy for us to make changes to the transmit power on the device. It is recommended that you reduce the transmit power to half and test it to see how the signal is received in the areas that it is required in. Again, there is no reason to transmit at full power unless you have a weak signal or dead spot somewhere that you are trying to overcome.

Defending from surveillance

The second step of the attacking methodology that we discussed in *Chapter 4, Breaching Wireless Security*, is the surveillance step. In this step, the attacker surveys and looks for a weak network to attack. In the example shown in the next screenshot, what kind of network do you think the attacker would attempt to attack?

```
CH 10 ][ Elapsed: 16 s ][ 2013-02-06 01:08

BSSID              PWR  Beacons    #Data, #/s  CH  MB   ENC   CIPHER AUTH ESSID

E0:91:F5:EC:21:0B   0      0         8    0   11  54e  WEP   WEP         GBCGMEN2.4
20:10:7A:13:D1:8B   0      4         0    0    1  54e  WPA2  CCMP   PSK  SBG658084
00:14:95:32:24:42   0     10         0    0    6  54 . WEP   WEP         2WIRE287
7C:BF:B1:72:38:B0   0      3         0    0    6  54e  WPA2  CCMP   PSK  ATT608
00:14:95:4C:7A:E2   0      9         0    0    6  54 . WEP   WEP         2WIRE228
00:25:9C:AB:BB:71   0      6         0    0    6  54e. WPA2  CCMP   PSK  MTB
E0:91:F5:DF:1D:A1   0      9         0    0   11  54e  WEP   WEP         VIGER
20:10:7A:12:9D:94   0     10         7    0   11  54e  WPA2  CCMP   PSK  Mordecai
E0:91:F5:BB:30:61   0     22         0    0   11  54e. WPA2  CCMP   PSK  WoodTigerLi]
00:14:AC:01:12:8C   0     22         2    0    6  54e. WPA   TKIP   PSK  TG852GC2
0E:14:AC:01:12:8C   0     25         8    0    6  54e. OPN              Kevin
00:24:56:CE:6C:41   0     16       124    0   10  11 . WEP   WEP         2WIRE417
0A:14:AC:01:12:8C   0     28         2    0    6  54e. WEP   WEP         Cardwell
20:10:7A:13:ED:56   0     25         0    0    6  54e  WPA2  CCMP   PSK  Castelli
06:14:AC:01:12:8C   0     24         1    0    6  54e. OPN              Amalia
00:1D:CF:B3:47:C0   0     25         0    0    6  54e  WPA2  CCMP   PSK  TG852GC2
38:6B:BB:00:5D:20   0      9         0    0    1  54e  WPA2  CCMP   PSK  ATT360
40:B7:F3:46:4C:20   0      9         0    0    1  54e  WPA2  CCMP   PSK  ATT328
20:E5:64:B1:A7:D0   0     19         0    0    1  54e  WPA2  CCMP   PSK  ATT048
90:B1:34:D1:3E:90   0      8         0    0    1  54e  WPA2  CCMP   PSK  ATT408
08:86:3B:15:6C:5E   0      3         0    0    1  54e  WPA2  CCMP   PSK  miles2822
2E:24:81:BB:D2:B6  -1      5         0    0    6  54 . OPN              hpsetup
```

While many of you would probably be tempted to go after the access points that use the WEP protocol, an attacker would usually go for the open access points that are not protected by any form of encryption. This is because if the access point is not protected, it is just a matter of connecting to it and then carrying out the rest of the process. This should provide you with the first step for defending your network once you have made the power modifications that are allowed by your device. The next step is to configure encryption on the access point. Since we are several chapters into the book at this time and have discussed the flaws of WEP, there should be no doubt in your mind that we need to select one of the WPA protection options for the access point.

Configuring encryption

The challenge we have is that there is no standard, and different access points will provide different methods and names for the settings. The options of the Bountiful router are shown in the following screenshot:

When you look at the previous screenshot, how do you know what it means? Well, this is where you either consult the manual or configure a particular setting and then see what happens. We all know that WEP is not an option, so we can eliminate that right away. The challenge is all of the variants of WPA. What exactly do they mean? The important thing to remember is that the WPA configuration changes from a key that is based on the hexadecimal numbering system to a password or passphrase. This means that, as a user, you have to select a strong password or passphrase. You may recall that we cracked a dictionary word that was used in a WPA configuration. So, one of the first things you want to do is think of a strong password or passphrase to use for your configuration. On this configuration page, we do not have much information to go on other than the options WPA/WPA2 and personal/enterprise. Which one should we use? The best way to think about this is that you want to use the strongest possible option that you can configure and set up easily.

The **WPA2 Enterprise** option usually requires a RADIUS server to act as the authentication server. Now that we have discussed some of the options, refer to the previous screenshot once again. Which one would you say is the strongest and best protection scheme? This is kind of a trick question because, as we have already stated, the strength of the authentication comes down to the password or passphrase. The recommendation is that you use a password or passphrase that is greater than 14 characters in length. In the Bountiful interface, we really cannot tell what algorithm is being used. There have been some attacks against WPA that were not classic dictionary attacks, so we do not want to select one of them.

In WPA, we have two main cryptographic algorithms. The first one, which has been around the longest, is referred to as WPA Personal or WPA PSK TKIP. The TKIP stands for Temporal Key Integrity Protocol; and since it has been around the longest, some attacks have been developed against it. In fact, it has not been recommended for wireless security since 2006. The second algorithm is AES that stands for **Advanced Encryption Standard**. This is the algorithm that is recommended today, and this is usually what is meant by WPA2 PSK Enterprise. If you refer to the next screenshot, you can see what the AES configuration page looks like in Kismet:

```
0
            Name: belkin.0a4
           BSSID: 08:86:3B:39:20:A4
           Manuf: BelkinIn
      First Seen: Feb  6 01:37:44
       Last Seen: Feb  6 01:37:44
            Type: Access Point (Managed/Infrastructure)
         Channel: 9
       Frequency: 2452 (9) - 1 packets, 100.00%

            SSID: belkin.0a4
          Length: 10
            Type: Beacon (advertising AP)
802.11d Country: TW
                 Channel 1-11 20dBm
      Encryption:  WPA PSK AES-CCM
```

The challenge is, if you refer to the following screenshot of the settings available in the Bountiful router, the router really does not come out and state AES. Fortunately, most of the newer routers will state whether the configuration setting is AES or not. An example of this is shown in the following screenshot:

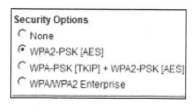

If the wireless router you have does not have AES listed, as shown in the previous screenshot, you can either consult the user manual or configure the router and then use one of the tools that we have covered to see if you can detect what algorithm is being used.

Settings that provide protection from masquerade and rogue AP attacks

This brings us to the masquerade and rogue AP attacks. The best defense against these types of attacks is encryption; by this we mean the encryption of the data on your network. This will prevent an attacker from successfully compromising WLAN traffic because it will be encrypted even if they manage to break in to the wireless network. This is best accomplished by setting up a Virtual Private Network (VPN). This is an essential component to deploy when you are connected to wireless hotspots. There are a number of VPN services out there, but before we get into them, we will look at examples with and without a VPN. If you refer to the next screenshot, you will see the network data from a wireless network that is not protected by a VPN.

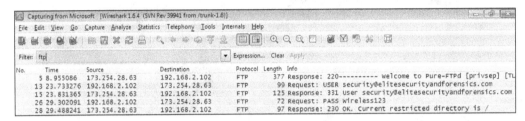

As you can see, the FTP protocol is clear text, and as such, the data that was entered (that is, the username and password) has been compromised on this network. The way to protect this is to use a VPN and encrypt the data. The next example shows the same network data after a VPN setup.

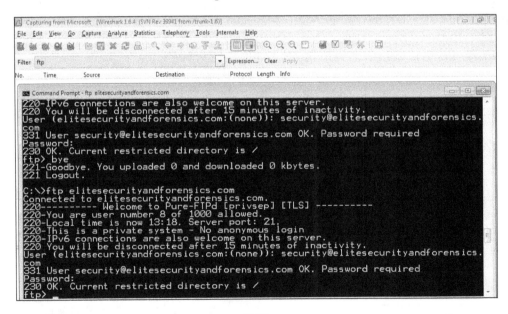

As you can see from the next screenshot, there is no information about the username and password; in fact, there is nothing on the protocol either! This is the power of a VPN and, to reiterate, it is essential that you use these when connecting from public hotspots and do everything from within a VPN. If you'll look at the example in the next screenshot, you will see that there is nothing in the network traffic other than the encrypted traffic; the network traffic is all-encrypted.

You are probably wondering how to get a VPN. Well, there are a number of them on the market, and it is beyond the scope of this book to cover them in detail. As with most software, there are reviews available online; so if you'll conduct a search in your favorite search engine, you will be able read the reviews for the different VPN products. There are three that we will mention here. The first one is the OpenVPN product; it has both free and commercial versions, and you can find out more about it at `http://openvpn.net/index.php/open-source.html`. The second one we will discuss is the proXPN product. For the PC, it is free; for mobile devices, it requires payment. You can find out more at `http://www.proxpn.com/`. The last one we will talk about is the BananaVPN. There is no free version of this; you can get more information about it at `http://www.bananavpn.com/`. It does not matter what product you choose, but it is essential that you get the VPN software and start protecting your data in transit.

The last protection mechanism we will cover is the firmware. It is essential that you update the firmware of your wireless devices. Just like software, these devices require updating. It is no different from patching your operating system or software on your computers or mobile devices. Unfortunately, it is often neglected, and as a result of this, many attacks have been conducted against the device itself. For an indication of this, you can go to www.routerpwn.com and see a list of attacks against a variety of different router manufacturers. This is just a small sample of the attacks out there, so update your firmware. Do it today!

Summary

In this chapter, we have discussed the measures you need to take to secure your wireless network. It is imperative that you select a data protection configuration to make your wireless network as secure as possible. The recommended setting is WPA2 with the Advanced Encryption Standard (AES). For those of you who want to go the extra mile and configure your wireless network to the level of an enterprise configuration, you might want to review the documentation at www.freeradius. org and experiment with building a RADIUS server to handle your authentication. It is not the easiest setup to deploy, but it does provide you with an enterprise-level wireless network. For home use, this is not required; the WPA2 AES is as strong as most will ever need. Bear in mind that you need to set a complex password with a minimum length of 14 characters. Refer to the following table for a small sample of how long it takes to crack a password depending on the password's length. For more information, refer to http://www.geodsoft.com/cgi-bin/crakcalc.pl.

Length of password	Estimated crack time (assuming 100 billion guesses per second)
Seven – alphanumeric and symbols. For example, HHffg!!.	5 minutes
Nine – alphanumeric and symbols. For example, HHffg!!68.	2.5 months
Fourteen – alphanumeric and symbols. For example, HHffg!!68340fg.	15 million centuries

What usually works best is a phrase that only you know. Finally, we looked at the reality that a wireless network is, in fact, open to the air, and the result of this is that the data is not protected in transit in most cases. Our solution for this is to deploy a VPN. It is imperative that we always use a VPN when connecting to public access points or hotspots.

In conclusion, there are things we can do to protect our wireless networks. The techniques we have discussed throughout the book will provide you with the best protection possible and allow you to continue to use and enjoy the conveniences of wireless networks in a safe and secure manner.

Wireless Tools

In this appendix, we will list a number of the tools that are available for wireless networks. We've used some of the tools throughout the book, while we've not used some others at all. The intent is to provide the readers with a one-stop reference for looking up the tools to support them in their quest for securing and learning more about wireless networks. The tools are not listed in any particular order.

- `aircrack-ng`: It is a powerful suite of tools that can be used to crack WEP/WPA and other security protocols. `aircrack-ng` is an entire distribution of tools to use when assessing wireless security. While the tool was originally written for the Linux platform, there is also a Windows version. As with most tools, the Windows version is not as robust; furthermore, it would require you to develop your own DLLs, and this is not something most would want to do. So, it is best to stick with the Linux version. The tools within `aircrack-ng` are very powerful and should be part of your wireless auditing toolbox. For more information, refer to `http://www.aircrack-ng.org`.

- `aireplay-ng`: It is used to inject/replay frames, and with attacks like the cracking of WEP, this can speed up the process. The `aireplay-ng` tool is used in combination with the `aircrack-ng` tool. The concept is to generate the packets and capture them, and then apply the captured traffic to the `aircrack-ng` tool. The `aireplay-ng` program is very powerful and can implement a large number of attacks. For more information, refer to `http://www.aircrack-ng.org/doku.php?id=aireplay-ng`.

- `airmon-ng`: It is used for placing a card in monitor mode, which turns off filtering on the network card, and in effect, allows all traffic to be received. Since the card is placed in monitor mode, it only displays 802.11 wireless network traffic. The `airmon-ng` utility will also display the status of interfaces if executed without any parameters. For more information, refer to `http://www.aircrack-ng.org/doku.php?id=airmon-ng`.

- `airodump-ng`: It is used as a protocol analyzer to capture raw 802.11 wireless packets. The tool is used in conjunction with the other `aircrack-ng` tools for testing wireless security. There is GPS capability, which provides the ability to log the coordinates of the access point (AP) that are found. For more information, refer to `http://www.aircrack-ng.org/doku.php?id=airodump-ng`.

- `airpwn`: It is mainly a hacking tool that allows you to attack wireless networks by eavesdropping transmitted packets between client and access point. The tool will listen and look for a specific pattern, and once it finds a match, it will carry out an attack, such as spoofing (pretending to be someone else). For more information, refer to `http://airpwn.sourceforge.net/Airpwn.html`.

- Kismet: It is an excellent scanning tool that can not only detect access points but that also has the capability to perform as an intrusion detection system (IDS). This capability is carried out using what is referred to as **kismet drones**. More information about this and other features can be found at `http://kismetwireless.net/documentation.shtml#readme`.

- `ssidsniff`: It is used to scan for access points and to capture and save wireless traffic to a file. The tool has a scripting capability that allows it to be customized and configured to meet the different requirements of an audit. For more information, see `http://www.monolith81.de/ssidsniff.html`.

- `dsniff`: It is a set of tools that can be used for a number of tasks; it can be used to identify protocols that are using clear text communication and to display the authentication credentials that are captured from the network. For more information, refer to `http://en.wikipedia.org/wiki/DSniff`.

- ettercap: It is a powerful tool that can be used as a sniffer and much more. The tool can perform man-in-the middle attacks (MiTM) and ARP poisoning and can display authentication information from network traffic. It has a scripting language that you can use to filter, modify, and inject data into network packets. The tool can also be used to intercept communications of encrypted protocols. ettercap has many features and is a tool at which you should take a look. For more information, go to `http://ettercap.github.io/ettercap`.

- inSSIDer: This tool is similar to Kismet, but it was started for the Windows platform. inSSIDer products are free, and there are also a number of commercial products available with enhanced features. The program allows you to scan for access points and display a number of parameters about each access point that is discovered. Features allow you to measure signal strength and identify the coverage of the signal to determine interference obstacles. For more information, go to `http://www.metageek.net/products/inssider`.

- Ekehau: It is a commercial site survey tool that shows the wireless coverage of access pints. The tool can be used to identify weak signal areas and improve wireless network design. Another feature of the tool is that you can upload a map, and the tool will map the access point signal strength with respect to the provided map. Ekehau has an add-on of a wireless spectrum analyzer that can be used in conjunction with the tool to perform a complete analysis of the wireless frequency spectrum. There is a suite of tools to review; for more information, see `http://www.ekahau.com/`.

- AirMagnet: It is a comprehensive software suite that provides numerous solutions across the wireless landscape. Some of the available solutions are as follows:
  - Wireless lifecycle
  - WLAN design
  - WLAN security and performance monitoring
  - WLAN analysis and maintenance

  For more information, refer to `http://www.flukenetworks.com/enterprise-network/wireless-network/AirMagnet-WiFi-Analyzer`.

- Cisco Spectrum Expert: It takes spectrum analysis to another level; not only does it scan the frequency spectrum, it also classifies the devices that it encounters. This allows the source of the interference to be localized as well as identified. For more information, go to `http://www.cisco.com/en/US/products/ps9393/index.html`.

- AirDefense: It provides a multitude of products for security and compliance. The product provides for wireless IDS. AirDefense eliminates the threat of rogue access points by analyzing the traffic and prioritizing threats to the network. For more information, see `http://www.airdefense.net/index.php`.

- Yellowjacket: It is a wireless receiver module designed to work with HP's iPAQ® PocketPC®. The receiver is used to analyze wireless channels and identify the information and characteristics of a wireless architecture. Yellowjacket is a mobile hardware platform that can be carried into the field for analysis. For more information, refer to `http://www.bvsystems.com/Products/WLAN/Yellowjacket/yellowjacket.htm`.

Index

Thank you for buying
BackTrack – Testing Wireless Network Security

About Packt Publishing

Packt, pronounced 'packed', published its first book *"Mastering phpMyAdmin for Effective MySQL Management"* in April 2004 and subsequently continued to specialize in publishing highly focused books on specific technologies and solutions.

Our books and publications share the experiences of your fellow IT professionals in adapting and customizing today's systems, applications, and frameworks. Our solution based books give you the knowledge and power to customize the software and technologies you're using to get the job done. Packt books are more specific and less general than the IT books you have seen in the past. Our unique business model allows us to bring you more focused information, giving you more of what you need to know, and less of what you don't.

Packt is a modern, yet unique publishing company, which focuses on producing quality, cutting-edge books for communities of developers, administrators, and newbies alike. For more information, please visit our website: www.packtpub.com.

About Packt Open Source

In 2010, Packt launched two new brands, Packt Open Source and Packt Enterprise, in order to continue its focus on specialization. This book is part of the Packt Open Source brand, home to books published on software built around Open Source licences, and offering information to anybody from advanced developers to budding web designers. The Open Source brand also runs Packt's Open Source Royalty Scheme, by which Packt gives a royalty to each Open Source project about whose software a book is sold.

Writing for Packt

We welcome all inquiries from people who are interested in authoring. Book proposals should be sent to author@packtpub.com. If your book idea is still at an early stage and you would like to discuss it first before writing a formal book proposal, contact us; one of our commissioning editors will get in touch with you.

We're not just looking for published authors; if you have strong technical skills but no writing experience, our experienced editors can help you develop a writing career, or simply get some additional reward for your expertise.

BackTrack 5 Cookbook

ISBN: 978-1-84951-738-6   Paperback: 296 pages

Over 80 recipes to execute many of the best known and little known penetration testing aspects of BackTrack 5

1. Learn to perform penetration tests with BackTrack 5

2. Nearly 100 recipes designed to teach penetration testing principles and build knowledge of BackTrack 5 Tools

3. Provides detailed step-by-step instructions on the usage of many of BackTrack's popular and not-so- popular tools

BackTrack 5 Wireless Penetration Testing Beginner's Guide

ISBN: 978-1-84951-558-0   Paperback: 220 pages

Master bleeding edge wireless testing techniques with BackTrack 5

1. Learn Wireless Penetration Testing with the most recent version of Backtrack

2. The first and only book that covers wireless testing with BackTrack

3. Concepts explained with step-by-step practical sessions and rich illustrations

Please check **www.PacktPub.com** for information on our titles

Advanced Penetration Testing for
Highly-Secured Environments:
The Ultimate Security Guide

Learn to perform professional penetration testing for highly-secured
environments with this intensive hands-on guide

Lee Allen

PACKT open source *

Advanced Penetration Testing for Highly-Secured Environments: The Ultimate Security Guide

ISBN: 978-1-84951-774-4   Paperback: 414 pages

Learn to perform professional penetration testing for highly-secured environments with this intensive hands-on guide

1. Learn how to perform an efficient, organized, and effective penetration test from start to finish

2. Gain hands-on penetration testing experience by building and testing a virtual lab environment that includes commonly found security measures such as IDS and firewalls

3. Take the challenge and perform a virtual penetration test against a fictional corporation from start to finish and then verify your results by walking through step-by-step solutions

Quick answers to common problems

Metasploit Penetration
Testing Cookbook

Over 70 recipes to master the most widely used penetration
testing framework

Abhinav Singh

[ ] open source

Metasploit Penetration Testing Cookbook

ISBN: 978-1-84951-742-3   Paperback: 268 pages

Over 70 recipes to master the most widely used penetration testing framework

1. More than 80 recipes/practical tasks that will escalate the reader's knowledge from beginner to an advanced level

2. Special focus on the latest operating systems, exploits, and penetration testing techniques

3. Detailed analysis of third party tools based on the Metasploit framework to enhance the penetration testing experience

Please check **www.PacktPub.com** for information on our titles

www.ingramcontent.com/pod-product-compliance
Lightning Source LLC
Chambersburg PA
CBHW082122070326
40690CB00049B/4186